QUALITY ASSURANCE OF SEAFOOD

Carmine Gorga
President, Polis-tics Inc.
Gloucester, Massachusetts

Louis J. Ronsivalli M.S., Retired,
Former Director
Gloucester Technological Laboratory
National Marine Fisheries Service
Lawrence, Massachusetts

An **avi** Book
Published by Van Nostrand Reinhold
New York

An AVI Book
(AVI is an imprint of Van Nostrand Reinhold)
Copyright ©1988 by Van Nostrand Reinhold
Library of Congress Catalog Card Number 88-12227
ISBN 0-442-23053-2

Printed in United States of America

Designed by Rose Delia Vasquez

Van Nostrand Reinhold
115 Fifth Avenue
New York, New York 10003

Van Nostrand Reinhold (International) Limited
11 New Fetter Lane
London EC4P 4EE, England

Van Nostrand Reinhold
480 La Trobe Street
Melbourne, Victoria 3000, Australia

Macmillan of Canada
Division of Canada Publishing Corporation
164 Commander Boulevard
Agincourt, Ontario M1S 3C7, Canada

16 15 14 13 12 11 10 9 8 7 6 5 4 3 2 1

Library of Congress Cataloging in Publication Data

Gorga, Carmine, 1935-
    Quality assurance of seafood / Carmine Gorga,
    Louis J. Ronsivalli.
        p. cm.
    "An AVI book."
    Bibliography: p.
    Includes index.
    ISBN 0-442-23053-2
    1. Fishery processing—Quality control. 2. Fishery
products—Quality control.
I. Ronsivalli, Louis J. II. Title.
SH335.5.Q35G67 1988
664′.9497–dc 19   88-12227
                                                                     CIP

Dedication

To the members of the U.S. seafood industry, one of our nation's oldest and most important industries, and especially to:

- The fishermen who often expose themselves to high degrees of physical risk and are faced with varying types of economic handicaps, especially in their struggle to compete against more heavily subsidized foreign fishermen;
- The spirited, pioneering seafood processors and retailers who have begun to take the necessary steps to ensure the quality of their seafood products to the consumer;
- The restaurateurs who, by their ability to assure the quality of the seafood that they serve, have lifted the image of the entire class of seafood from the lowest to the highest eating experiences;
- The dedicated and resourceful scientists, epitomized by the late Dr. John T.R. Nickerson of M.I.T., whose efforts will eventually enable more consumers to obtain high-quality seafood at retail outlets.

Contents

Preface

This book is based on a U.S. government research project that studied the quality assurance of seafood and its impact on the economics of processing.

Quality assurance, which is determined at the point of consumption, serves a twofold function: it increases consumer demand and eliminates losses due to spoilage. This book outlines a plan that will increase the domestic supply of seafood. We include recommendations for some administrative revisions and a description of four target areas for improvement in fisheries production in the United States.

* Accelerated growth in aquaculture
* Development of fisheries
* Preservation of temporary oversupplies
* Improvements in processing efficiency

Seafood is a vital and growing source of protein in the U.S. food supply, and is potentially a multibillion-dollar industry. For these reasons, it is in the best interests of the United States to develop its seafood productivity.

Added costs to assure quality should not be construed as a deterrent to achieving that goal. On the contrary, we subscribe to a paradox proposed in the *Harvard Business Review* of July/August 1986 by Professor Wickham Skinner, professor of business administration at the Harvard Business School: "When low cost is the goal, quality often gets lost. But when quality is the goal, lower costs do usually follow."

The resolution to this paradox is that when quality is assured, the substantial losses due to product spoilage and the adverse effect on demand caused by customer dissatisfaction are greatly lessened. When fixed costs are distributed over a large production base, the cost per unit will be lower than for those costs distributed over a small production base. Finally, less effort is required to sell a product that has a positive image in the mind of the consumer.

It will be noted that the text places emphasis on fish and fish fillets and virtually omits reference to shellfish and processed fish products (canned, cured, pickled, or otherwise treated). The reason is that fresh and frozen fish products, and especially fish fillets, have most often been cited for lack of good quality by investigative teams that have surveyed them at the marketplace, particularly in retail supermarkets; these are the products that have given negative connotations to the word *fish*. However, the quality of shellfish products at the retail level is usually high, partly because some of the shellfish are marketed in the live state (such as lobsters, oysters, clams, and crabs), and partly because their high value motivates handlers to avoid spoilage losses. There seem to be no indications of lack of quality in canned and other processed seafood products. Still, the observations and recommendations made in the text are applicable to the handling of all seafood.

This book is directed at a broad audience: the seafood industry; government agencies (federal, state, and local); the academic community; the general public; politicians; nutritionists; dieticians; food editors; and students of food science, nutrition, and dietetics. For this reason, we have attempted to present the material in nontechnical language. At the same time appropriate scientific discourse has been included to give interested scientists the technical rationale behind some of the statements and conclusions reached. Parts of the text that may be less relevant to the general audience but useful to scientifically trained readers are set apart in appendices, although even these are not in the strictest technical style, because not all readers are equally trained in all fields of food science.

In the interest of time, and because some of the discussions may not be of immediate interest to everyone, the text has been so organized that individual members of the seafood industry do not have to read the entire book to obtain the information they need. Still, the text contains much information that is of at least ancillary interest to all, and we recommend that the book be read in its entirety. It provides insight into the reasons why there is a seemingly more rapid spoilage of seafood than there is of meats; why frozen seafood products are apparently of poorer quality than unfrozen ones; why seafood bought in the supermarkets often is not of as good quality as that bought in specialty stores; why the U.S. seafood industry is much more important than is generally perceived; and why seafood is so much more important than other protein sources.

Finally, our lengthy and broad experience has brought us to the conclusion that the study of seafood provides a variety of scientific challenges worthy of sophisticated academic and research efforts. These efforts, for which there is a reasonably high probability of success, should produce gratifying rewards to those who seek the opportunity to make important scientific and technological contributions to the literature. At the same time these researchers would help the United States attain the high economic and health benefits that can be derived from one of its most valued, natural, renewable resources.

Acknowledgments

Although we cannot name each of the innumerable contributors to this book, we would like to express our sincerest thanks to all concerned, especially James Bordinaro, Sr., founder of the Empire Fish Company, Inc.; Charles Shackleford, plant manager of the Great Atlantic and Pacific Tea Co., Inc.; Ronald H. Carignan of the DeMoulas Super Markets, Inc.; Gus Aslanis, president of the Aslanis Fisheries; Anthony Amoriggi, Jr., plant manager of Amoriggi Brothers Seafoods; the late Salvatore J. Favazza, executive secretary of the Gloucester Fisheries Commission; Jake Dykstra of the New England Fisheries Steering Committee; John D. Kaylor, Joseph H. Carver, Joseph M. Mendelsohn, and Burton L. Tinker of the Gloucester Laboratory of NMFS; and finally, Thomas J. Moreau, Philip J. McKay, and Vernon Rix of the NMFS Northeast Inspection Office.

PART

1

ASSURANCE OF SEAFOOD SUPPLY

CHAPTER

1

The Need to Assure
the Seafood Supply

FOOD IN HISTORY

It is universally accepted that man's basic needs are food, clothing, and
shelter. This short list does not include such items as water, oxygen,
and sun—perhaps because, although vital, in their natural state ordinarily they are not marketable commodities; nor does the list include
innumerable social and cultural needs. Of man's three basic needs, food
is not only the most important, it is the only vital one. People cannot
survive for long without food, but they can without clothing and shelter.

In the beginning of his existence, man was forced to spend most of his
time and effort simply obtaining the food required for his sustenance.
He derived his food from edible plants, animals, fish, birds, amphib-
ians, reptiles, and insects. He was nomadic, following herds and moving on from areas where the supply of edible plants had been spent.
Thus he had little or no time to engage in other activities and it is sur-
mised that, for all practical purposes, it took all of the efforts of one
person to supply the food required by one person.

If we look at this relationship as an indication of food-supplying
efficiency, we obtain a value that will be useful later: a ratio of 1:1.
To determine food-supplying efficiency, we will use a simple fraction
in which the numerator represents the total number of people to be fed,

and the denominator represents the number of people engaged in the activity that supplies the food. Namely, $FSE = N/W$ where FSE stands for food supply efficiency, N the number of people to be fed, and W the number of workers engaged in supplying the food.

But man was not a loner. It is speculated that the groups to which early man belonged were relatively small, numbering about twenty members in all. Division of labor began with the formation of these groups, allowing everyone to participate in one of the several activities that were aimed at food procurement. The most able-bodied were the hunters of animals; the others gathered plant foods and perhaps prepared the hunting tools for the hunters. Thus, in essence, it took the output of the entire group to supply the food required by the group.

Man's food-gathering activities in the earliest days resembled those of animals, but his superior intelligence eventually permitted him to outstrip animals. According to evolutionary theory, man's full-time mobility on two limbs was largely influenced by his ability to use his forelimbs to wield a club or to throw rocks and, later, spears in order to enable him to capture animals that were faster and/or stronger than he. His ordinary animal abilities enabled him to learn to avoid poisonous or otherwise harmful foods, but it was his superior intelligence that eventually enabled him to surmise the role of seeds in plant reproduction.

This latter development enabled man to grow plants, making them available without having to search for them. Once man recognized the benefits of assuring survival and quality of life by purposefully planting seeds, he made a quantum leap to the tactics that assured his food supply. This activity eventually marked the beginning of non-nomadic, sedentary civilization.

The Impact of Agriculture on Man

The development of agriculture affected man's entire life. The convenience of permanent shelters became evident, and this led to the establishment of hamlets, which in turn developed into villages. Soon there emerged the concept of boundaries to protect the planted fields, and many battles must have been fought over these. No doubt the outcome of many of these battles depended on the food supplies of the armies involved. The art of war itself may have gotten its beginnings from man's food-gathering activities, for what else would have been used in combat except the weapons—such as clubs, rocks, spears, and slings— that had already been devised to hunt the larger animals?

The same ingenuity that man had applied to devise hunting weapons was eventually devoted to the development of tools and equipment to operate farms, to cut trees, and to build houses. At a much later stage, when farming reached the point that one farmer could produce enough food for two people, one person was freed to engage in activities other than those concerned with the procurement of food. Those demonstrating special skills or interests would gradually become engaged in producing tools or other goods and providing services, such as grain milling, storage, transportation of the crops, and even trade. Products reached ever-expanding markets, and goods produced in one area became available to people who, living in other areas, might have never seen those items before. Trade also permitted man to expand the variety of foods in his diet and to use ever-increasing amounts of spices and condiments. Indeed, these very activities motivated explorations, land acquisitions, and the building of empires.

As development followed upon development, farming efficiency increased, and the food produced by one farmer could feed more people than it had in the past. Ever more people were freed to undertake other pursuits, causing mushrooming developments in every area of human endeavor. As a result, the human race was permitted time for leisure, recreation, and the pursuit of a large variety of other "nonproductive" activities.

The practice of agriculture and trade has taught us the value and wise use of water to irrigate the land and to provide navigation and energy. Unfortunately, it has also brought us to activities that have resulted in the extinction of many species of animals and in the pollution of the environment.

From a nutritional point of view, an important aspect of the development of agriculture is that it converted man from dependence on a predominantly meat diet to dependence on a predominantly grain and vegetable diet.

THE NUTRITIONAL COMPOSITION OF FOODS

Of the major nutritional elements in foods, two are of special importance in seafood: proteins and lipids. Although there are also other components (see figure 1.1), they are not of great relevance for the purposes of the present discussion, and by concentrating our attention on proteins and lipids, we can better show the dietary superiority and possible therapeutic value of seafood. These characteristics are outlined in figure 1.2 and discussed in some detail below.

Proteins

Proteins play a unique and varied role in our diet. They are required for the building and rebuilding of body tissues, the conduct of body

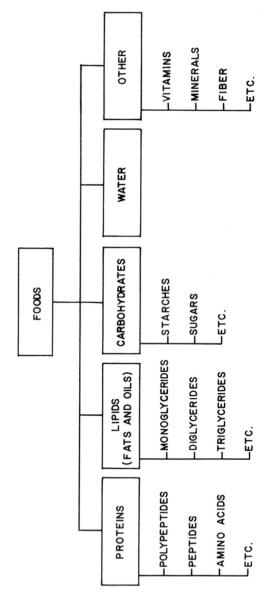

Figure 1.1. Nutritional Components of Foods.

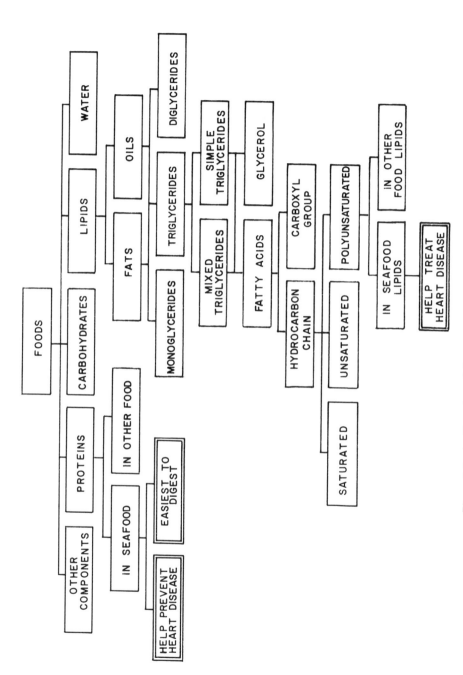

Figure 1.2. An Overview of the Dietary Superiority of Seafood.

7

chemistry, the defense of the body against disease, and a host of other activities. We cannot survive unless we are able to ingest an adequate supply of proteins. Amino acids form the primary structures of all proteins.

Of the amino acids found in proteins, ten are classified as essential—arginine, histidine, isoleucine, leucine, lysine, methionine, phenylalanine, threonine, tryptophane, and valine—the remaining, nonessential. Man requires both types for metabolism and growth. The chief difference between essential and nonessential amino acids is that our bodies cannot synthesize the essential ones or at least cannot synthesize them in sufficient quantities. Therefore, since the essential amino acids must be obtained from outside sources, their constant supply must be maintained through the diet.

Whether people eat flesh foods or plant foods to satisfy their protein requirements may be a matter of choice or it may be dictated by taste, custom, religion, or a code of ethics. Whereas protein foods vary in their amino acid composition, it is well known that meats, fish, poultry, eggs, and milk are good sources of the complete array of essential amino acids. Accordingly, even though early man did not possess a scientific knowledge of his nutritional requirements, by virtue of the fact that he was basically a carnivore, he managed to satisfy his needs for the essential amino acids and other nutrients.

We may also obtain protein from plant foods, especially the seeds of plants, however, the protein in any given plant is not complete. In order to assure the ingestion of all of the essential amino acids in a vegetarian diet, it is necessary to eat plant foods in particular combinations. But there is good reason to believe that even an optimally balanced vegetable diet, although theoretically adequate, cannot provide the same nutritional value as the protein derived from animal sources. It is therefore recommended by some nutritionists that at least one-fourth to one-half of protein intake be from animal sources.

Lipids

The lipids include substances that are commonly called fats and oils. The oils to which we refer are those found in biological systems (plants and animals) and not those derived from petroleum or shale; the essential oils (mostly terpenoids) occurring in plants also remain outside the scope of our discussion.

In general, when a lipid is in the solid state at room temperature, it is called a fat; when in the fluid state at room temperature, it is called an oil. Examples of the former are lard and butter, and of the latter, vegetable oils, such as corn oil, and the fish oils, such as cod liver oil. There are two characteristics that account for the difference in their physical state at room temperature: the length of the hydrocarbon chain in the fatty acids and the degree of unsaturation within the chain. These two characteristics are explored in greater detail in appendix 1.

THE NUTRITIONAL VALUE OF SEAFOOD

Very likely, fish has been used for food by man even before the earliest recordings of history. In man's earliest attempts at recorded expression, there are clear pieces of evidence of the use of fish for human consumption. However, until recent years, no matter how valuable fish might have been to man, its value would never be anything other than a means to satisfy hunger. With the current knowledge of the impact of the diet's components on our health and with the knowledge of the composition of foods, however, we may select fish as food because we know that its value is beyond just that of an adequate source of protein.

The Superiority of Seafood as a Protein Source

Of all of our sources of protein, seafood is recommended for those whose lives are threatened by cardiovascular disease. Whereas the meats of most land animals are relatively high in fats, those of most seafood are relatively low in fats. Whereas the meats of most land animals are associated with a high cholesterol content, those of most seafood are associated with a low cholesterol content. And whereas the lipids in the meats of land animals are formed from a relatively high proportion of saturated fatty acids, the lipids in seafood are formed from a relatively high proportion of unsaturated and polyunsaturated fatty acids. People who suffer from atherosclerosis are especially encouraged to choose meats from lean fish over those from fatty fish, although this is a debatable issue, as will be seen below (see table 1.1).

In widespread efforts to shed light on the causes of atherosclerosis, one study has linked the low incidence of heart attacks among Greenland Eskimos with their unusually high consumption of fish (Bang et al 1980). This discovery sparked a number of studies (Glomset 1985) on the composition of fish lipids and has resulted in some convincing conclusions that thus far have been corroborated rather than disputed. All of the available evidence regarding the structure and the chemistry of fatty acids strongly indicates that seafood is the only food we consume in which the degree of polyunsaturation of some of the fatty acids is high enough to provide a favorable effect in relation to atherosclerosis. Thus, a major aspect of the nutritional value of seafood is emerging that further removes it from the role of just an alternative protein source.

Table 1.1 Dietary Features of Various Protein Foods (per serving)

Feature	Beef	Cheese	Egg	Lamb	Plants*	Pork	Poultry	Seafood*
1. Completeness of protein	X	X	X	X		X	X	X
2. Source of vitamins and minerals	X	X	X	X	X**	X	X	X
3. Relatively low in calories							X**	X†
4. Relatively low in fats					X**		X**	X†
5. Relatively low in cholesterol					X		X**	X
6. Relatively high in polyunsaturated fats					X			X
7. Relatively high in calcium		X	X		X			
8. Recommended in diets for those having cardiovascular disease							X**	X†
9. Possible therapeutic value in treatment of certain cardiovascular diseases								X†

*Each of these classes comprises a wide variety of species that may contain a wide range of nutrients. The plants include grains, nuts, and beans and the products made from them.
**Applies to only some species or only for specific cuts or parts.
†Applies to many species.

Although polyunsaturated fatty acids can be found in both plants and seafood, the degree of polyunsaturation is higher in fish oils than in plant oils. The fatty acids in plant oils are polyunsaturated to the degree that they exhibit more than one site of unsaturation, with most of them containing two. They rarely have three sites and almost never have more than three. Fish oils, on the other hand, have a much higher degree of polyunsaturation, with some of their fatty acids showing as many as six sites of unsaturation. It should also be noted that the lengths of the hydrocarbon chains of some of the fatty acids in fish oils are significantly longer than those found in the fatty acids of vegetable oils.

A number of studies have been undertaken to try to uncover the reason for the apparent correlation between high seafood consumption rates and low heart attack rates (Kromhout et al 1985). One thrust of the experiments has been to monitor the effects of adding fatty acids from fish oils to the diet in order to determine their effect on the clotting tendency of blood (Dyerberg et al 1978). One fatty acid identified as having a dramatic mitigating effect on the clotting characteristic of blood is eicosapentaenoic acid. This is a long chain, polyunsaturated fatty acid having twenty carbon atoms and five sites of unsaturation. One theory is that the anti-atherosclerotic effect of this acid is a result of its ability to influence a change in the nature of the blood platelets so that it reduces their tendency to adhere. In normal conditions, under the influence of thromboxane (a powerful clotting agent produced in the body), blood platelets enable the blood to clot as a means of stopping bleeding from minor cuts and abrasions. Under certain abnormal conditions, the clotting tendency of the platelets may reach excessive proportions, resulting in a build-up of platelets on the inner linings of blood vessels. One effect of this condition is that it may lead to the formation of blood clots that can then block vital arteries to the heart, causing heart attack, or to the brain, causing stroke. Another effect of this condition is believed to initiate the accumulation of lipids on the vessel walls with their subsequent degeneration.

Under the influence of eicosapentaenoic acid, the body's production of thromboxane is impeded, and in its stead a compound is produced that is similar to thromboxane but has only a small fraction of the clotting influence of that agent. With this alteration, the adhering tendency in the platelets is lowered, and as a consequence, the tendency to block arteries and to form clots is also lowered. This alteration of the clotting ability of the blood, however, cannot be carried to an extreme; otherwise, the tendency to bleed may be raised to levels that would make bleeding from minor cuts and abrasions difficult to stop. Thus, any attempt to alter the clotting tendency of the blood platelets must be controlled in order to achieve a healthy balance that minimizes both a dangerous clotting and a dangerous bleeding tendency.

Numerous other claims are being investigated concerning the therapeutic value of eicosapentaenoic acid (EPA) as well as docosahexaenoic acid (DHA), the two predominant omega-3 fatty acids found in seafood. They are also reported to protect the body against migraine headaches, rheumatoid arthritis, some cancerous tumors, mature-onset diabetes, some allergies, and even morning stiffness and tender joints. Since some of these claims are gradually being validated, it is only natural that manufacturing corporations became interested in extracting those oils and producing fish-oil pills which are already in the market.

Vitamin and Mineral Value of Seafood

Most seafood represents an excellent source of vitamins and minerals for human nutrition, and some is also an abundant source of calcium, which has a preventive effect on osteoporosis.

Digestive Value of Seafood

Of all the sources of protein, seafood is the easiest to digest. It does not contain indigestible elements such as cellulose or fiber, which exist in vegetable diets, nor cartilage (gristle) or tendons, which are encountered in most meat diets. Hence seafood is recommended for people who must stay on especially delicate diets.

Weight-Control Value of Seafood

One additional value of seafood concerns the general weight-control aspects of foods. Although a few meats (e.g., white meat of poultry) are as low in fat as most seafood, and although some seafood (e.g., mackerel, trout, herring) is relatively high in fat content, seafood is generally lower in calories than other proteins are and thus has an overall superiority over other protein foods in weight-control diets.

Value of Seafood for Medicinal and Industrial Purposes

Many extracts of aquatic animals have only begun to be studied for medicinal and industrial uses. Aequorin, a phosphorescent substance extracted from one type of jellyfish, has been used to measure the most minute changes in the calcium concentration in body fluids and cells. Extracts from other coelenterates have been used for the treatment of high blood pressure. Prostaglandins, a class of hormones occurring in only very small amounts in mammals, have been found in abundance in sea whips. This means that medical research concerning the application of these compounds for the treatment of a variety of human ailments may proceed unhampered and at accelerated rates.

One potent neurotoxin isolated from a marine worm has been found to be a powerful insecticide, and a synthetic modification of it was originally marketed in Japan as the insecticide Padan. Tetrodotoxin, another powerful neurotoxin, occurring in puffer fish, has the effect of blocking the flow of nerve impulses and has been used to minimize the pain in patients afflicted with terminal cancer. Extracts from a wide variety of mollusc have exhibited antiviral, antibacterial, or antitumor activities. Extracts from sharks have yielded heparinlike compounds that appear to have five times the effectiveness of commercial heparin, without its adverse side effects. Carrageenan prolongs the activity of commonly used analgesic and antitussive agents such as codeine and ethylmorphine. Broad-spectrum antibiotics have been reportedly extracted from marine sponges.

The whole subject of the use of aquatic plants as foods or as sources of pharmaceuticals and other industrial products has been purposely omitted, only because sea plants are not of significant economic importance to the United States at this time. In South and Southeast Asia, however, they have been used as human food, and in Europe, as animal feed. The reportedly poor digestibility and closely guarded production technology of seaweeds have minimized the use of these products which otherwise appear to have a good potential for future use (Anon. 1983, and Bruce 1983).

One area of significant economic potential, as mentioned above, can be found in the recovery of the large amounts of oil as a by-product of fish meal production, and the preparation of fish-oil capsules by health product makers. As reported by Hollie (1986), the U.S. market for omega-3 pills is expected to top $200 million in 1988 from about $30 million in 1986.

THE ECONOMIC EFFICIENCY OF SEAFOOD PRODUCTION

There is yet another aspect to the superiority of seafood as a protein source and that is its high economic efficiency of production as compared with other sources of animal proteins. Using closed systems of aquaculture, fish and shellfish have the highest protein production potential for a given growing area of any other animal produced as food. Current estimates are that as much as one million pounds of seafood could be grown per acre per year in closed-system silos. Also, in accordance with current information, fish are the most efficient converters of animal feed, requiring about one and a half pounds of feed to produce one pound of fish. Other animals used for human consumption require larger amounts. The ratio for beef, for example, is at the other end of the spectrum, 10:1; namely, it takes about ten pounds of feed for every pound of animal produced.

REFERENCES

Anon. 1983. Seaweeds—products and markets. *Infofish Marketing Digest.* (4) 23–26.

Bang, H. O., J. Dyerberg, and H. M. Sinclair. 1980. The composition of the Eskimo food in northwestern Greenland. *Am. Jnl. Clin. Nutr.* 33: 2657–2661.

Brisson, G. J. 1981. *Lipids in Human Nutrition.* Fort Lee, NJ: Jack K. Burgess.

Bruce, C. 1983. Seaweed as food. *Infofish Marketing Digest.* (4). 27–29.

Dyerberg, J., H. O. Bang, E. Stoffersen, S. Moncada, and J. R. Vane. 1978. Eicosapentaenoic acid and prevention of thrombosis and atherosclerosis. *Lancet.* 2: 117–119.

Glomset, J. A. 1985. Fish, fatty acids, and human health. *N. Engl. J. Med.* 312(19):1253–1254.

Hollie, P. G. 1986. Fish oil makes a splash. *The New York Times*, Dec. 7, 1986, p. 1F.

Kromhout, D., E. B. Bosschieter, and C. de Lezenne-Coulander. 1985. The inverse relation between fish consumption and 20-year mortality from coronary heart disease. *N. Engl. J. Med.* 312(19):1205–1209.

LaBuza, T. P. 1977. *Food and Your Well-Being.* St. Paul, MN: West Publishing.

Nettleton, J. 1985. *Seafood Nutrition: Facts, Issues and Marketing of Nutrition in Fish and Shellfish.* Huntington, NY: Osprey Books.

Tannahill, R. 1974. *Food in History.* New York, NY: Stein and Day.

Tannenbaum, W. R. 1977. *Food Proteins.* Westport, CT: AVI Publishing Co.

2

Strategies for Assuring
the Seafood Supply

The value of seafood to man, as outlined in chapter 1, is increasingly becoming a matter of common knowledge throughout the world. Not surprisingly, as a consequence of the barrage of information being distributed through scientific journals, as well as trade and even fashion magazines, the demand for high-quality seafood is increasing rather steadily. Per capita consumption in the United States is reported in table 2.1. However, before addressing directly the issues of what specifically is a "high-quality seafood" and, indeed, how one can assure the quality of seafood to the consumer, it is useful to look at the issue of the future supply of these commodities.

The basic question is, Are there enough supplies of seafood to meet the increasing demand? At one extreme is the simple reality that without supplies, it is academic to be concerned with issues of quality; at the other extreme is the direct and simple fact that assuring the quality of seafood means preserving its quality and ultimately preserving the very stock of the existing supply.

NATIONAL AND INTERNATIONAL SUPPLIES

In addressing the issue of the need for an increase in the supplies of seafood, the concern is twofold: one is with the availability of supplies worldwide, the other is with that of each specific country. As can be

Table 2.1   U.S. Per Capita Consumption of
Fish and Shellfish

Year	Pounds*
1910	11.2
1920	11.8
1930	10.2
1940	11.0
1950	11.8
1960	10.3
1970	11.8
1975	12.2
1976	12.9
1977	12.7
1978	13.4
1979	13.0
1980	12.8
1981	12.9
1982	12.3
1983	13.1
1984	13.6
1985	14.5

*Edible meat.
Source: U.S. Department of Commerce, 1986,
Fisheries of the United States, 1985

seen in table 2.2, there has been a phenomenal growth in world
landings. The very rate of exploitation of the worldwide biomass during
this century has led many to be concerned about the possible exhaus-
tion of the stocks. Many marine species, as is well known, are on the
endangered list.

The United States has consistently occupied the fourth or fifth position in world landings during the twentieth century. As can be seen
from table 2.3 and figure 2.1, at times there are dips in American
commercial landings for human consumption (there are also "industrial" landings—fish destined to industrial uses as fertilizers, fish meal,
and the like—as well as the catch of sportfishermen, but these are not
included in this table). Every time such dips occur, and these dips are
much more accentuated when they are reported for smaller geographic
subdivisions, the fear arises that the supply of seafood—or at least the
supply of specific species of fish or shellfish—might come to an end.

The concern for stabilizing, and eventually steadily increasing, the
current supply of seafood is therefore entirely warranted. Within each

Table 2.2 U.S. and World Commercial Fishery Catches (in million metric tons)

Year	United States Weight excluding mollusc shells*	United States Weight including mollusc shells**	World
1950	2.2	2.6	21.1
1960	2.2	2.8	40.2
1970	2.2	2.8	65.6
1975	2.2	2.8	66.4
1976	2.4	3.0	69.8
1977	2.4	3.0	68.9
1978	2.7	3.4	70.4
1979	2.8	3.5	71.1
1980	2.9	3.6	72.0
1981	2.7	3.8	74.8
1982	2.9	4.0	76.6
1983	2.9	4.1	76.8
1984	2.8	4.8	82.8

*Published by U.S. Department of Commerce.
**Published by Food and Agriculture Organization of the United Nations.
Source: U.S. Department of Commerce, 1986. *Fisheries of the United States*, 1985.

Table 2.3 U.S. Commercial Landings of Fish and Shellfish (landings for human food)

Year	Million Pounds
1975	2,465
1976	2,775
1977	2,952
1978	3,177
1979	3,318
1980	3,654
1981	3,547
1982	3,285
1983	3,238
1984	3,320
1985	3,294

Source: U.S. Department of Commerce, 1986, *Fisheries of the United States*, 1985.

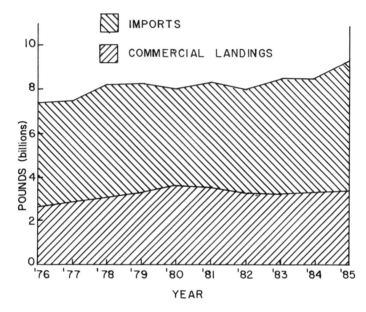

Figure 2.1. U.S. Supply of Edible Fishery Products, Round Weight, 1976–1985.

nation so inclined, however, can this concern be transformed into a national mission? We will show that there is much that can and ought to be done to achieve this goal.

RELIANCE ON IMPORTS

At the limit, a nation (as distinguished from the world as a whole) can supply all of its needs for fishery products through imports. As can be seen in table 2.4, the United States is in fact importing fishery products to meet a large part of its seafood needs. Interestingly enough, however, the United States also has a significant export trade (table 2.5), but this aspect of the issues will be discussed in the last chapter. Here we are concerned with the sizeable increase of imports during the recent past, as represented in figure 2.1. Even though the numbers are given in *round weight* (weight of whole fish as caught) to make them equivalent to the domestic landings, while in effect imports are wholly edible, it becomes exceedingly clear that the current U.S. course of action of major reliance on imports is not advisable, nor is it for any country. Even landlocked nations should be encouraged to develop their fish productivity through inland fisheries and aquaculture because of the importance of an assured seafood supply.

Table 2.4   U.S. Imports of Edible Fishery Products

Year	Million Pounds
1975	1,913
1976	2,228
1977	2,176
1978	2,411
1979	2,359
1980	2,145
1981	2,272
1982	2,225
1983	2,387
1984	2,454
1985	2,754

Source: U.S. Department of Commerce, 1986.
Fisheries of the United States, 1985.

If there is a surplus capacity in the use of the fishing fleet, as is the case at present in most of the United States (i.e., some vessels are not fishing at all and others are reduced to part-time), a national policy that fosters an increase of imports would doom to idleness much existing capital and the manpower that accompanies it. Among other effects, a domestic supply achieved at increasing costs could be expected to spur potentially large increases in the price of fishery products.

Table 2.5   U.S. Exports of Edible Fishery Products

Year	Million Pounds
1975	218
1976	241
1977	331
1978	448
1979	554
1980	574
1981	669
1982	657
1983	602
1984	574
1985	648

Source: U.S. Department of Commerce, 1986
Fisheries of the United States, 1985.

Three more negative aspects of the reliance on imports for meeting the seafood needs of a nation strengthen the argument that this is a wholly undesirable policy. The first is the economic peril of an escalating trade deficit for any nation. The deficit due to trade imbalance in seafood is quite high for the United States. It was $5.6 billion in 1985 (U.S. Dept. Commerce 1986). The second aspect is the danger that, with the possible inability of the world's seafood supply to keep pace with the growing demand, the competition among buying nations could create serious procurement problems for any single nation. The third is that, as increasing worldwide awareness of the need to assure the quality of seafood is translated into reality, the rate of increase in world demand could expand even further. This situation may become more serious as the information regarding the superiority of seafood as a protein source becomes more widespread, tending thereby to further increase its worldwide demand.

The high probability that any single nation will not be able to obtain sufficient imports to meet its future seafood needs, even if it is able to pay, is reason enough to consider what has to be done in order to increase domestic production.

A FIVE-POINT PROPOSAL

There are at least five goals that, pursued together, would form a mission to gain short- as well as long-term solutions to the problem of assuring the availability of high-quality seafood supplies to the consumer.

1. The growth of aquaculture
2. Fisheries development
3. Preservation of temporary oversupply
4. Improvement of the efficiency of processing
5. Quality assurance

The Growth of Aquaculture

Aquaculture is one of the most fundamental and ancient methods of assuring the supply of high-quality seafood. Historians agree that it was practiced in one form or another for longer than can be accurately determined. There is evidence that the Chinese, as early as four thousand years ago, then the Greeks and the Romans, maintained fish ponds to secure a consistent supply of fish; and some authors have reported evidence, dated as early as that of the Chinese, of Egyptians

growing stocks of tilapia in ponds. During the Middle Ages, the maintenance of fish ponds had spread throughout Europe and parts of Asia.
In the United States most of the earliest attempts to develop aquaculture were largely aimed at the maintenance of hatcheries for stocking
purposes.

Today aquaculture is practiced throughout the world. Many species,
including abalones, buffalofish, carp, catfish, clams, crabs, crawfish,
eels, lobsters, milkfish, mullet, mussels, oysters, plaice, pompano,
salmon, shad, shrimp, sole, striped bass, tilapia, and trout, as well as
a few aquatic plants, are raised in systems that are partly or completely
controlled by man. Yet when we consider the sources from which we
derive our fishery products, we think first of the oceans, and then of
lakes, ponds, and rivers (freshwater fish are not strictly seafood because
they are not harvested from the sea). It is only as an afterthought that
even experts acknowledge the existence of artificial catfish ponds, trout
farms, and other forms of aquaculture. Even though the situation is
fast changing, aquaculture in the United States is still in its infancy.
One estimate is that aquaculture contributed only 3 percent of American consumption in 1975 and only 12 percent today (Kleinfield 1987).

We attempt to satisfy our requirements for seafood in the same ways
as we did centuries ago, when in fact there is a long overdue need for
the United States to embark on a coordinated and determined large-
scale effort to adopt aquacultural methods for producing fish and shellfish in major quantities. For the near future, an output of 20 percent
of the national need (current domestic landings plus imports for food
consumption) or about 1.7 billion pounds per year should not be an
unreasonable goal.

It is difficult to estimate the potential impact on American economy of
the proposed large-scale aquaculture effort, but there is every reason to
believe that it can be very large. The advantages of producing seafood
by aquaculture rather than by prevailing fishing methods include but
are not limited to the following:

1. There is much less risk of injury or death to the personnel employed
   in harvesting the fish and shellfish.
2. The total environment in which the fish and shellfish grow, including
   such water conditions as temperature and mineral content, can be
   controlled to the desired levels so that contaminants and disease can
   be avoided and growth can be optimized.
3. Habits, life processes, and traits of fish and shellfish can be studied in
   order to permit the possible hybridization and cultivation of species
   having the more desirable traits (e.g., faster growing, or disease
   resistant).

4. The feeding of fish, both in its composition and amount, can be controlled to achieve maximum results at least possible cost.
5. The rate of harvest is not dependent on weather conditions, one of the major limiting factors in traditional harvesting methods.
6. The harvest is proportional to the effort. There is no guessing as to the size of the expected catch or the quantity of other components of the harvesting efforts, the amount of ice, for instance.
7. The size of the stock can be controlled because any serious threat of its depletion is precluded, and therefore the market demand can be met with more certainty.
8. The time between the slaughter of the animals and their entrance into the handling chain is reduced from days (and possibly a week or more) to minutes. This promptness constitutes a quantum leap over traditional methods in terms of quality assurance and is reflected in higher levels of initial quality.
9. Seafood entering the distribution chain through these methods and processes can then compete with other meats in relation to shelf life.

All of these advantages result in two outstanding effects: The price of seafood has a greater chance of being stabilized, and the supply to the consumer can be increased to keep pace with the demand.

Fisheries Development

This method attacks the problem of possible shortages of supplies, and it seeks not only to increase the current supply of seafood, as aquaculture does, but also to reduce the pressure on those species that are threatened with extinction because of overfishing or any other cause.

The problem of insufficient stocks at sea is a long-standing one in the United States. From time to time, when the problem has become severe enough, the successful marketing of unpopular species has given the fishermen additional stocks to exploit. Lobsters, halibut, haddock, and pollock are only a few of the species that were once considered *trash fish* (fish for which no market exists) but that are now among the most highly valued and in popular demand. Although trash fish represent one of the most evident sources of a number of underutilized species that can eventually be made popular, there are still many other places where one might go to find species that can be added to the current seafood supply: One simply has to think of places deeper than or farther away from the traditional fishing grounds. Indeed, the very bulk of the biomass, the plankton, may be considered for the production of edible products.

But how does the introduction of a new species into the market generally take place? Traditionally, it has been proven that the introduction of new species in the market requires education of both the industry and the consumer; and often it also requires some form of government assistance in the field of basic research and technology.

The government generally considers the development of a particular species for a new fishery from several technical and economic viewpoints. For example, does the species under consideration resemble a popular one so that it would be readily accepted by consumers? Is the stock of the species of sufficient size to support a fishery? Is the species not utilized because of some undesirable or even adverse characteristic? Can the identified problem or problems be resolved? Studies are carried out from many points of view in order to discover solutions to problems connected with each species. In each case, the overriding concerns are to increase the supply of seafood to the consumer and to reduce pressure on endangered species.

Cooperative efforts by both government and industry in developing fisheries or in saving popular species under heavy fishing pressure or threat of severe depletion have often resulted in the addition of several important species to the seafood supply, with high benefits-to-cost (B/C) ratios. The following are a few cases involving Atlantic species, even though the government has been involved in the development of species from other areas as well.

The Northern Shrimp Fishery. Industry was unable to develop the northern shrimp fishery for export because it could not produce shrimp of a quality acceptable to foreign buyers. Once government scientists were able to develop a technique that would block or slow down the action of enzymes immediately upon harvesting, American fishermen were able to produce a product that easily satisfied the requirements of foreign buyers, thus firmly establishing the northern shrimp fishery (Lane 1973). In 1984, landings of North Atlantic shrimp were about seven million pounds (net weight). The benefits-to-cost ratio calculated for this effort was about 2000+ (for every dollar of tax payers' money spent, there was a return of more than $2,000 to the nation). The plus sign signifies that the number continues to grow, because in the fraction that yields the value of the benefits-to-cost ratio, the benefits continue to accrue with time whereas the cost is a one-time expenditure that does not increase. The assumptions used to obtain these B/C ratios are discussed in chapter 13.

The Red Crab Fishery. Because of high demand, the meats of lobster and crab tend to command high prices on the market. This demand also creates an intense pressure on their naturally limited stocks. The government tried to redress this imbalance between supply and demand by investigating the possibility of introducing into the market the red crab, an underutilized species whose meats appeared to have all of the organoleptic attributes of a substitute for lobster and crab. An earlier industry attempt to establish this fishery had failed, and when government scientists studied the case, they concluded that two impediments existed.

The first impediment, inefficiency of extraction of the meats, was overcome with the development, testing, and introduction of a modified mechanized processing system that was originally designed for processing snow crabs. This innovation made the processing of red crabs economically feasible. The second impediment was overcome through the sectioning of the crabs at sea as a method for landing them with the highest possible yield (unless the crabs are processed immediately, they shed their legs and render the harvesting effort uneconomical). These two measures were basic to the establishment of the fishery (Learson et al 1976). In 1984, landings were somewhat less than twenty million pounds (net weight). The benefits-to-cost ratio associated with this effort was estimated at more than 500. It should be noted that this estimate includes the benefits from the technology transfer resulting from the adoption by some blue crab processors of the newly redesigned machinery used to establish the red crab fishery.

The Pollock Fishery. The pollock is a gadoid species of the North Atlantic and in most respects is similar to haddock, although its meat is darker and, to some consumers, slightly less desirable in texture. From all other aspects, however, government technologists considered it a good candidate in their search for species that might satisfy the market demand for haddock, cod, and related finfish, which are in short supply.

A study of the issues revealed that the only impediment to the establishment of a pollock fishery was the lack of a moderate effort to educate industry and consumers. Accordingly, the effort was successfully undertaken by the government in close cooperation with the industry, and the pollock fishery was firmly established (Kaylor and Murphy 1970).

Within less than a year, the value of pollock was elevated from about $0.05 per pound landed weight to about $0.40 per pound landed weight; its current market value is nearly as high as that of other gadoid species. In 1984, landings of pollock were about 65 million pounds. The benefits-to-cost ratio for this effort has been estimated at about 2000 + .

The Ocean Quahog Fishery. With the demand for surf clams outstripping the supply, government scientists studied the possibility of introducing the ocean quahog (mahogany clam) in the market in order to help fill the demand for, and ease the fishing pressure on, surf clams (Mendelsohn et al 1970). The endeavor appeared especially feasible because of the estimated large stocks available, however attempts to use this species were beset by three major problems: The meats of the ocean quahog were reputed to be too tough to be useful, they were much darker than those of the surf clams, and their taste was described as having an intolerable "iodine" or medicinal tang.

The problem of toughness was readily resolved by exposing the meats to controlled heating under pressure; this treatment can be controlled to attain any tenderness level that is desired. The strong taste was found to be water soluble and, in diluted form, was discovered to be a concentrated form of the typical clam flavor. It was therefore partially leached out in some clam preparations and diluted in others. Indeed the strong taste turned out to be an advantage because in some preparations a desired clam flavor is attainable with a smaller amount of ocean quahog than other clams. The color problem was apparently of minor consequence because once the tenderness and taste problems were resolved, it did not hinder the marketability of this species.

As a consequence of government help, the ocean quahog was soon marketed in sufficient quantities to establish its popular use, and this fishery has become a permanent reality. In 1984, landings were about 40 million pounds (net weight) and the benefits-to-cost ratio associated with this effort was estimated at about 1000+.

In a few short years, the new fisheries discussed above have added about 132 million pounds per year to the domestic seafood supply, and these examples are only a fraction of the overall endeavor. Thus, it seems fair to predict that through similar efforts with other underutilized species, similar results might be possible, and in a relatively short span of time, through fisheries development, at least 750 million pounds per year could be added to the domestic seafood supply.

In addition to the introduction of new species into the market, there is the potentially large contribution of product development to be considered. New products can be derived from new species—as well as from waste recovery systems, from unusable species, and from combinations of fish and other meats.

Preservation of Temporary Oversupply

For some species there is a time of year when, because of the ready availability of stocks, quantities landed exceed demand, resulting in a glut in the market. At such times, the price is apt to drop so sharply that it hardly covers the costs expended in harvesting. Consequently, there is likely to be less than full use of the catches, and some of the product may eventually become spoiled. More frequently, the catches of most species are at a relatively low volume so that there is an insufficient supply to satisfy the demand. The value of the product goes up and results in relatively high prices to the consumer. The fluctuation in the market price of a given species due to this randomness of landings can be quite high—as much as 100 percent—and is disconcerting to everyone, particularly to the consumer.

What is of direct interest to us here, however, quite apart from the issue of price, is the fluctuation in quantities landed. It is likely that during times of glut, some of the fish will not be utilized to its economic potential. The solution to this problem is evident. If the industry could be encouraged to freeze and store a part of the catch in times of glut, that amount could be added to the seafood supply to help fill the demand during periods of scarce landings. Yet very few processors or middlemen have followed this course, while others give several reasons for the lack of interest in this solution.

The first reason is that frozen fish is not as valuable as fresh fish, and therefore, with the lingering hope that all or nearly all of the catches can be channeled into the fresh fish market, the surplus is not frozen. By the time the surplus can no longer be sold as fresh fish, it is usually too late to freeze it because, by then, it will have deteriorated in quality. The second reason is that some processors do in fact freeze low-quality fish, and some supermarket retailers mishandle even high-quality frozen fish. In so doing, these operators depress the market price of frozen fish and help spread the impression that all such fish, unavoidably, is of poor quality. As a consequence of these market conditions, the myth exists that frozen fish is not as good as fresh fish—simply because the fish is frozen.

The implication is that fish should not be frozen if its high quality is to be retained. This myth was challenged in at least three cases in which processors, working in cooperation with members of the National Marine Fisheries Service (NMFS) utilization research laboratories, built frozen-storage warehouses expressly for the purpose of freezing excess supplies. The warehouses were, and still are, maintained at about $-28.9°$ C ($-20°$ F). Since these processors froze the fish while it was still of high quality and kept it at adequately low temperatures, the quality of the frozen products later released to the market was high. In each case, in fact, the processors who participated in the projects were able to realize rewarding profits from their investments, and what is more important, these fish, which were preserved during times of glut and were channeled into the market during times of scarcity, greatly increased the supply available to consumers.

No reliable estimates exist on the potential amount of fish that might be preserved in this way; since temporary excess supplies occur for a number of species, however, it seems to be considerable. Conservatively assuming that these supplies are 5 percent of domestic landings for food consumption, in a relatively short span of time about 165 million pounds per year can be added to the domestic seafood supply. Quite apart from specific amounts, however, the general points are clear: following the proper procedures—which we shall examine in Part 2—insures that the consumer consistently receives high quality products and that the amount of supplies available increases.

Improvement of the Efficiency of Processing

In the United States, the conventional methods of processing fish and shellfish result in relatively large amounts of waste. For example, when fillets are cut from fish, approximately 35 percent of the original weight of the fish is recovered as food; the other 65 percent either is sold to producers of fish meal and fertilizers for a few pennies per pound or is discarded altogether. When the meats are separated from their shells, the recovery rate of edible shellfish meat is about 50 percent lower than that of fish fillets. The financial loss involved in these operations varies; in all cases, there is the cost of labor and other fixed costs involved in producing worthless rather than valuable products (loss of potential cost reduction). It might involve the cost of the price differential between what could have been recovered for human consumption and the value of fish meal (loss of potential earnings); and when the product is discarded altogether, there is also a net loss brought about by handling and transportation costs for its disposal (direct expenses).

The study of methods and processes to recover these losses by minimizing the processing waste is relatively recent in the United States, although it is advanced in countries such as Japan and in meat and poultry industries. It is only during the past few decades that deboning machines, for instance, have been introduced to the seafood industry, increasing the yield for both finfish and shellfish. Many developments, however, have taken place under the leadership and with the direct involvement of the federal government through the utilization research laboratories of the NMFS, particularly those in Seattle, Washington, and in Gloucester, Massachusetts. The NMFS is a division of the National Oceanic and Atmospheric Administration (NOAA), which is a division of the U.S. Department of Commerce (USDC). The NMFS conducts research in management of fisheries, and utilization of marine products. It enforces the regulations governing fishing within the jurisdictional boundaries of the United States, operates a voluntary seafood inspection program, protects endangered marine species, provides financial and technical assistance to commercial fishermen, and represents the United States in a number of international trade agreements.

The government's role in the preliminary studies and tests concerning waste recovery efforts was largely confined to the development of methods for increasing the efficiency of recovering meats from a few selected finfish species and from there it branched out into three related fields: the determination of the characteristics of the recovered product; its potential market use; and the potential applicability of the recovery method to shellfish and other finfish species.

An effective waste recovery system for the processing of crabs was thus developed. As a result, the Atlantic red crab made the transition from an underutilized to a highly acceptable species and is now in such demand that management of its stocks has become necessary to prevent its depletion. A market was also developed for simulated blue crab lump meat and other products.

When government scientists tested the feasibility of adding minced fish to either hamburger meat or frankfurters, they found not only that the addition of fish did not detract from the organoleptic qualities of these products, but that it enhanced their nutritional value. Indeed, the modification resulted in a significant reduction of the high cholesterol and total fat content usually found in these fast-food meats (King 1973 and Steinberg 1975). This potential use of the recovered product is significant because hamburgers and frankfurters are especially sought out by the young, and a high intake of cholesterol and total fat, even in youth, may have a lasting and adverse health effect.

As a result of this government effort, a number of deboning machines have been added to the standard processing equipment in a few plants, but their use is still not widespread. It is estimated that approximately 500 million pounds of edible meats can potentially be recovered and added to the current U.S. seafood supply if these machines are adopted by the industry.

Quality Assurance

Some of the fish that are landed and processed are eventually discarded because regardless of initial quality, they are allowed to reach such low levels of quality that they are unfit for human consumption. This end result may occur even when quality control measures are in place and are applied at various stages in the production and distribution process. It is estimated that losses due to spoilage are about 10 percent of the total catch. Indeed, our experience leads us to suggest that the figure is significantly higher. This provides the best evidence of the need to have not only quality control measures but also a quality assurance program in place.

The essential point here is that if all the fish and shellfish that are currently landed were of high enough quality to last up to the moment of consumption, there would be more than a 10 percent increase in supply (domestic landings plus imports), or about 850 million pounds per year.

For decades, academic and government research teams from many countries have conducted studies on the spoilage of seafood, methods of assessing quality levels, and methods for slowing the rate of spoilage, although the information garnered from these studies was never systematically applied. During the 1970s, however, the U.S. government and some members of the industry cooperated on a project that, with the application of quality assurance measures, resulted in a demonstrated and lasting improvement in the quality of seafood available in supermarkets—the place where spoilage occurs with greatest consistency and most deleterious effects, since some of the spoiled fish is liable to be sold to the consumer (Gorga et al 1978).

Part 2 of this book is devoted to quality assurance, which is the only one of the five proposed goals for increasing the supply of seafood for which a convincing body of data has been gathered. It has the potential to give the highest return in the shortest time and is basic to the successful application of all other proposed methods of increasing domestic productivity. Furthermore, it is the only comprehensive approach: Its concern starts at the very moment of preparation for the next catch and then follows the fish thus caught, up to the moment of consumption.

LESSONS LEARNED FROM EXPERIENCE

In aquaculture, where the payoff is almost assuredly large but the investment is of corresponding size and very long-range, satisfactory developments are occurring only for a few species, catfish, trout, and mussels being perhaps the most shining examples. There is, however, relatively little commercial production in this country concerned with such promising species as shrimp, for example.

In fisheries development, the situation is only slightly different. Although there is generally little financial risk, the process is slow and may not succeed. Significant developments have taken place in this field in the United States, partly because the introduction of new species into the market can at times be successfully carried out by private industry with little or no government support and partly because the government has applied the necessary effort to produce considerable results.

In fully utilizing temporary excess supplies, one finds conditions that are akin to those of fisheries development. Hence, a few success stories are encountered, but there is no continuing and comprehensive effort underway.

In improving the efficiency of processing, the payoff is similar to that of aquaculture: It is almost assuredly large, but the investment is correspondingly great and what is worse, very long-range. Therefore, there are only a few areas in which satisfactory developments are occurring. The recovery of crabmeat through the use of deboning machines is perhaps the most important example.

The same overall condition prevails in the field of quality assurance. Whereas the industry has not been able to do all that is necessary by itself to assure the quality of its products to the supermarket consumer, with the guidance of the federal government it has demonstrated beyond any doubt its willingness and capacity to work in a coordinated fashion and ultimately to upgrade the quality of its products. Indeed, fitfully aided by newcomers, a few firms are persevering along the path shown by the cooperative government/industry project that demonstrated the economic and technological feasibility of the necessary procedures to attain that goal (Gorga et al 1978). Although at times the number of participants has increased, there was little or no coordinated effort. Hence, those who do try to elevate the overall quality of the products available to the supermarket consumer tend to struggle more than they perhaps should. Producing a high-quality product costs more in the short run, but benefits generally are reaped in the long run. A quality assurance program is more risky than most because profits depend on the actions of the entire chain just as much as on one's own efforts. This risk is reduced as the degree of intra-industry coordination increases.

What is the general conclusion, then, that one is almost compelled to make? It is clear that the five goals discussed above make up the elements of a mission that, if faithfully pursued, has the attainable potential for high economic gain for the United States. And yet, can industry be expected to take on the mission, even if it is convinced of its validity? As logic dictates, and experience has demonstrated, none of the steps will be undertaken by industry without the leadership and actual involvement by the federal government.

REFERENCES

Anders, F. S. Jr. 1971. Artificial cultivation. In *Our Changing Fisheries*. Shapiro, S. (Editor). Washington, DC: Superintendent of Documents.

Brown, E. E. 1983. *World Fish Farming: Cultivation and Economics*, 2nd ed. Westport, CT: AVI Publishing Co.

Gorga, C., J. D. Kaylor, J. H. Carver, J. M. Mendelsohn, and L. J. Ronsivalli, May 1978. The Technological and Economic Feasibility of Assuring Grade A Quality of Seafoods. In-house report of the Gloucester Laboratory.

Kaylor, J. D., and E. J. Murphy. 1970. On the creation of the pollock fishery. Personal communication. Gloucester, MA: NMFS Gloucester Laboratory.

King, F. J., and J. H. Carver. 1970. How to use nearly all of the ocean foods. *Commercial Fisheries Review* 32(12):12–19.

King, F. J., J. H. Carver, and R. Prewitt. 1971. Machines for recovery of fish flesh from bones. *American Fish Farmer* 2(11):17–21.

King. F. J. 1973. Acceptability of main dishes (entrees) based on mixtures of ground beef with ground fish from under-used sources. *J. Milk and Fd. Technol.* 36:504–08.

Kleinfield, N. R. 1987. Across America, the fish are jumpin': It's a bigger business than chicken. Even tilapia and skate wings sell. *The New York Times*, March 1, pp. 1F, 8F.

Lane, J. P. 1973. On the creation of the northern shrimp fishery. Personal communication. Gloucester, MA: NMFS Gloucester Laboratory.

Learson, R. J., B. L. Tinker, V. G. Ampola, and K. A. Wilhelm. 1976. Roller extraction of crabmeat. *Proceedings of 1st Annual Tropical and Subtropical Fisheries Technological Conference.* Vol. II, Oct.

Mendelsohn, J. M., P. S. Parker, E. D. McRae, F. J. King, and A. H. Joyce. 1970. The ocean quahog—a bountiful clam. *Fd. Prd. Develop.* Nov.

Power, E. A., and W. L. Peck. 1971. The National Picture. In *Our Changing Fisheries.* Shapiro, S. (Editor). Washington, DC: Superintendent of Documents.

Ronsivalli, L. J. April 27, 1981. Benefits to U.S.A. of operating the Gloucester Laboratory during 1960–1980. In-house report of the Gloucester Laboratory.

Ronsivalli, L. J., J. D. Kaylor, J. P. McKay, and C. Gorga. 1981. The impact of the assurance of high quality of seafoods at point of sale. *Mar. Fish. Rev.* 43(2):22–24.

Steinberg, M. A. 1976. On the feasibility of substituting 10% fish flesh for beef in the production of hamburger and frankfurters. Personal communication. Seattle, WA: NMFS Seattle Laboratory.

U. S. Department of Commerce. 1981. *Fisheries of the United States, 1980.* Washington, DC: NOAA, NMFS, U.S. Department of Commerce.

U. S. Department of Commerce. 1985. *Fisheries of the United States, 1984.* Washington, DC: NOAA, NMFS, U.S. Department of Commerce.

U. S. Department of Commerce. 1986. *Fisheries of the United States, 1985.* Washington, DC: NOAA, NMFS, U.S. Department of Commerce.

3

The Role of Government in Assuring Seafood Supply

In a perfect society, the role of the government can be restricted to the administration of defense, justice, and the money supply. Indeed, there are some who would entrust even the administration of the money supply to private hands.

But society is far from perfect, and government has often found the need to play a role in assuring both the supply and the quality of the seafood available to the consumer. Let us briefly see how and why governments, both here and abroad, have performed this function.

A HISTORY OF THE GOVERNMENT'S ROLE

A study of the animal kingdom reveals that, in general, there are two areas of behavior that are predominant and that spur animals to extreme physical feats and risks. The first is the procurement of food and its protection from others. The second encompasses the propagation of the species. Psychologists may argue which of the two is predominant, but no one will argue that high priorities are indeed assigned to the procurement of food. The available historical information, as briefly reviewed in chapter 1, indicates that man was a social animal from the very beginning and that obtaining food was largely a group effort. It should also be stressed that in the earliest times the procurement of food was time consuming and that the number of people within each group engaged in the occupation of obtaining the food necessary to feed the entire group was high.

In most cases, adult males hunted for animals, while adult females and children foraged for edible plants. If all of the members were thus involved, the food supply efficiency (FSE) for the group would be assigned the number one ($FSE = N/W$ where N/W = number of consumers/number of suppliers). However, it is more likely that any given group would contain one or more children too small to help forage for food and one or more adults incapacitated or otherwise unable to participate in the food procurement activity. In such cases the food supply efficiency for the group would be higher, out of necessity; but the effort made by each individual in supplying the food would necessarily have to be intensified or the amount of food allotted to each would have to be reduced.

It can readily be appreciated that the activity of supplying food eventually reached a point that it required a high degree of planning and organization, which man, with his superior intelligence, had the capability to employ. As the social groups grew in size in the millennia that followed, the efficiency of procuring food was probably increased in small increments, but activities directed at supplying food continued to hold a high priority.

When the size of the social groups grew to such large proportions that rulers over each group were established, the rulers assumed the responsibility of assuring the supply of food. They also stored grains and other foodstuffs for any emergency, and in historical accounts of the Roman, Greek, Egyptian, and earlier civilizations, we find much evidence of this practice. Today, as in all of history, the supply of food is considered of such high importance that it warrants the highest priority in public affairs. Very early, the U.S. government made its intention in this regard quite clear. In the motion that elevated the position of the head of the USDA to a cabinet post, it was stated: "Agriculture, the single most important economic activity in the Nation, should be represented in the innermost councils of Government" (Stefferud 1962, p. 7).

Role of U.S. Government to Assure U.S. Food Supply

The U.S. government has done more to assure the food supply to its people and to millions of citizens of other countries than perhaps any other government in history (see figure 3.1). Other governments have done their share, to be sure, but none has approached the agricultural achievements of the United States. In 1862 under the administration of President Abraham Lincoln, the United States, then a relatively young

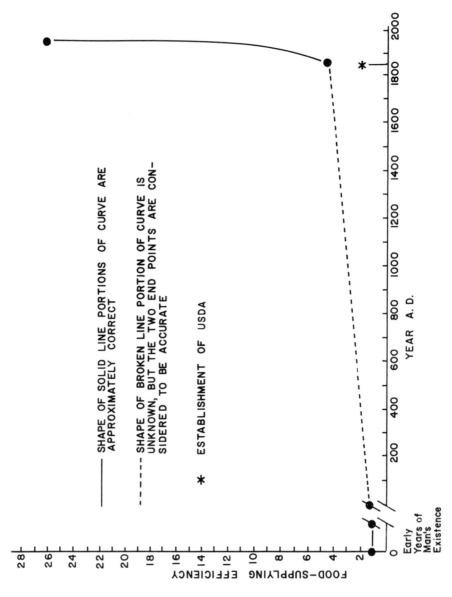

Figure 3.1. The USDA Impact on Man's Food Supply Efficiency.

34

nation that was into the second year of a civil war, enacted a law establishing the United States Department of Agriculture, now known widely as USDA. It is reported that this action was motivated by the successes achieved by the English, who for two decades had operated such an agency in an effort to assist farmers to enhance their productivity. Later that same year, two additional bills, the Homestead Act and the Land-Grant College Act, were passed. They both contributed powerfully to the development of our agricultural potential. From the beginning, the USDA worked in cooperation with farmers and with the Land-Grant Colleges on the solutions to problems that affected the production of grains, vegetables, fruits, food animals, and other agricultural products. Later, this was expanded to include the basic sciences and engineering, and to such applied fields as forestry, animal husbandry, veterinary medicine, statistical analysis, quality control, and applied engineering.

In 1889, the position of USDA commissioner was elevated to that of secretary in the cabinet, giving the department the highest possible ranking within the government structure. By 1938, the USDA had grown to about 100,000 employees and had recorded numerous significant achievements in both basic and applied areas of science and engineering. By then the department also realized the need to emphasize its efforts in utilization-oriented activities and, consequently, established its four Regional Utilization Laboratories: the Eastern Regional in Wyndmoor, PA, Northern Regional in Peoria, IL, Southern Regional in New Orleans, LA, and Western Regional in Albany, CA.

On the whole, within a mere hundred years, the agricultural advances made possible by the USDA, working in full cooperation with the nation's farmers, the Land-Grant Colleges, and, later, the State Agricultural Experiment Stations, resulted in outstanding contributions to the United States and to man in general.

As an overall appraisal of these accomplishments, it might be pointed out that whereas the food-supply efficiency of man took eons to increase from 1 in the beginning to about 4.5 in 1862, within just a hundred years (1862–1962), the efficiency of the United States farmers was increased to about 26.

As its food supply efficiency increased, the United States released more people to pursue other fields which, in the view of the authors, permitted an expansion in the many activities that propelled the United States into a position of leadership in both the industrial and agricultural revolutions. The scientific and engineering advances that were made by the United States in this relatively short time transformed it from a fledgling nation to the largest food producing country, the leading industrial nation, and the world's strongest political power. These outstanding successes and the prodigious benefits accrued to the United States are perhaps the best justification of the role performed by the U.S. government in the development of the agricultural resources of the nation.

THE LAG IN THE DEVELOPMENT OF FISHERIES RESOURCES

Projected against the phenomenal development of agricultural resources, the lag in the development of those of U.S. fisheries becomes clearly manifest. During the nineteenth century and the early part of this century, the United States is reported to have landed more than an eighth of the world's annual catch. In 1880, for instance, domestic landings for human consumption were a little over 1.7 billion pounds. Even though in absolute terms the amount of landings in the United States has increased (in 1985 they stood at about 3.3 billion pounds), its proportion of the world's catch has declined. In 1968, this figure had been reduced to one-twenty-fifth. According to the latest available figures, it has risen slightly and now stands at about one-twentieth, a small improvement, but an improvement nonetheless. And yet, this is still much less than during the last century; and clearly much less than it could be.

Why has the development of fisheries resources lagged behind those of agriculture? Before we attempt to give an answer to this question, let us dispose of the possible objection that the question itself is ill-advised—possibly because the decline has taken place in relative rather than absolute terms or even because there is a limit after all to natural resources. This objection is not valid because, as we have seen in the preceding chapters, a great deal can, and ought to, be done to increase the supply of domestically produced seafood.

A probing of the reasons for the lag yields at least three sets of answers: the first is economic; the second, technological; and the third, administrative.

Economic Reasons

To examine the full set of economic reasons the development of fisheries resources in the United States has lagged behind that of agricultural resources is largely beyond the scope of this book. But in addition to pointing to the impact of macroeconomic policies, one more answer can be given here. It is related to the set of questions involved in any project that requires a large-scale research effort before it can be implemented.

Quite apart from the issue of costs involved in any duplication of efforts, one that must be taken into special consideration when the point of view is that of the nation as a whole, there are a variety of factors that determine the economic feasibility of private research efforts. One has to start with the very size of the effort and continue with the size of the market to which the effort is directed, passing through issues of timing and freedom of entry into the eventual market. Laws, starting with patent laws, are also relevant factors.

What would be the logic of, say, even a large seafood processor's expending large amounts of funds to research and develop a market product that cannot be protected by patent rights since the product itself is a natural resource? One would then simply fall into a competitive cost disadvantage in relation to most other processors. Technologically, this is the essential nature of much of the research and development effort that is still needed to exploit fisheries resources.

Technological Reasons

The question of particular interest here is: Should the government assist in the development of fishery resources or should it be an exclusively private enterprise?

Aquaculture. In the previous chapter a case was made for a major thrust in aquaculture because our seafood supply will never be assured without it. And yet, it must be recognized that aquaculture will not be a major economic force if left to industry because it is not a simple undertaking. Just as it took the combined efforts of several teams of USDA professionals in the applied sciences to develop agriculture to its present state, so it will take a number of professionals in the applied sciences to develop aquaculture to a state that resembles our production of beef, pork, lamb, and poultry. This is an undertaking of such scope that it is beyond the financial and research capabilities of the seafood industry—just as the development of agriculture was, and would be even today, beyond the reach of the nation's farmers. There is no experience that can be reported for the U.S. government's involvement in aquaculture of the scope that we are advocating. Substantial, though isolated, cases of support have been related mostly to hatcheries.

Fisheries Development. The introduction of new species into the market does at times occur without any government intervention. Yet the process is slow and fraught with the danger of doing irreparable damage in the meantime to those species that are currently accepted and whose stocks might be depleted. Just the problem of communicating the right information from consumers to retailers, or from fishery biologists to fishermen, are enormous. The industry is scattered over vast distances and has no recognizable degree of integration. And communication is perhaps one of the more simple elements among the necessary factors that make development possible.

Preservation of Temporary Excess Supplies. The cost of building and maintaining freezer-warehouses is considerable, and most processors dealing in fresh fish are rather small economic operators for whom the building of the necessary facilities is usually out of the question.

Improving the Efficiency of Processing. Two aspects of the deboning machine technology mentioned in the previous chapter require consideration in order to understand the logic of the industry's reluctance to adopt those methods without the government's involvement, even though the amount of product that could be recovered by the processors is considerable. The first is that these methods require additional processing equipment and working space as well as modifications in the current processing procedures, changes that are not readily accepted by the industry without some assurance of economic advantage. The second aspect of seafood waste recovery systems is concerned with the nature and the marketing of the recovered product. The various machines used to recover waste material are built on a principle that reduces the product to the form of hamburger. Whereas the wide acceptance of this type of meat makes it a valuable commodity, the demand for ground fish is limited. Consequently, why would a seafood processor buy the equipment and pay for space and labor to produce a product for which, at first, there is only a limited market? Even though it may be possible for large processors to invest in a marketing program, what is known about the product's organoleptic and storage characteristics to enable them to market the product? Without a specific research program, there is no way of knowing whether minced fish is as easily converted to marketable products as hamburger is.

In fact, in one case in which members of the private sector made such an attempt, it was a costly failure because the action was premature. When it appeared that there might be a market for frozen blocks made of minced fish flesh, the industry produced a large amount of these blocks before their storage and use characteristics were adequately researched. Most of this product was not used and had to be destroyed because of adverse organoleptic changes (especially toughening and discoloration) that developed before it could be consumed. Thus, the lessons learned from experience in this area clearly show that private industry does not have the basic assets to assimilate waste recovery systems without government assistance.

Yet when the industry's progress was held to a pace commensurate with the government's findings and recommendations, the result was a permanent assimilation of the waste recovery systems by those few companies that worked in cooperation with the government.

Quality Assurance. As we shall see in part 2, a quality assurance program requires a high degree of coordination of activities from the moment the fish are caught to the moment they are bought by the consumer. Thus, fishermen, processors, and retailers must cooperate in many respects.

Fishermen perform relatively unique functions, considering the personal and economic risks that are associated with their trade. But whatever else they do, they conduct a business as well. As with other businessmen, the money they receive for their effort must be adequate to cover the total cost of operations, including their wages and a reasonable profit to provide the incentive to make them continue in their line of work. In fact, the price that fishermen receive for their catch should cover not only costs, wages, and profits for each fishing trip, but it should also cover the losses associated with trips in which part or all of the catch is lost, those in which the catch is so small as to represent a real operating loss, and a host of other economic losses, including those related to vessels and gear. Clearly, these are the expected risks in the fishing operation. What, then, is the incentive for fishermen to increase their costs in order to upgrade the quality of the fish landed, as required by the quality assurance program, unless they have a near certainty that increased costs can be recovered upon selling the product?

One of the added costs to the fisherman results from the strict maintenance of the temperature of the product. In most cases it means that more ice has to be taken for each fishing trip; in some cases it means that new equipment, such as reusable, washable, tote boxes, will have to be bought. In still others, trips have to be shortened, thereby raising the cost by an increased proportion of the travel time to and from the fishing areas. The average fuel and other vessel costs per trip and the average man hours per trip that are not associated with fishing are thus increased. There will, at times, be a larger proportion of the fishermen's time to stow the fish because of the greater attention required to conform to correct stowage (e.g., the extra care required to ensure correct layer packing of the fish with the ice on board the vessel) and there will also be a need to improve the handling of the product aboard the vessel in terms of sanitation. Often this extra care involves the expenditure of money for necessary modifications and equipment. These costs are recovered only if the processor pays a higher price for the fish landed. But then the question is shifted down along the marketing chain.

What is the incentive for processors to pay fishermen a higher price, unless they too recover the added cost? Processors will also have increased expenses in order to maintain the required temperatures in their plants, and while some will need special equipment, all will face increased energy costs; they will all have to pay additional labor costs to perform the required inspection and to meet the sanitary requirements.

Finally, retailers also will have to pay a higher price for the product and spend money to attain the required low temperatures in storage rooms and display cases. In addition, they will spend more money for the labor required to maintain a reliable FIFO (first in, first out) system and an effective system of inspection. And what is the assurance that retailers will recover their added costs?

The answer to these questions is evident: All members of the industry will recover their added costs only if the consumer pays a higher price for the final product. But consumers must receive something in return: not only a product of high quality, but a product of consistently high quality.

Given these facts, how will the consumer be assured of getting high-quality seafood? The only way out of these apparently conflicting but actually complementary needs lies in a coordinated program of action. In fact, there is no question that, once a quality assurance program is in place, the added costs will be recovered easily. Because the processor would not be allowed to introduce poor-quality fish into the program, fishermen would face competition only from those who land high-quality fish. The same is true for processors, middlemen, and retailers. As a result, consumers would readily pay a higher price for fish whose quality was most assuredly high.

A Partial Answer. On the basis of the evidence gathered from experience in quality assurance, fisheries development, aquaculture, or any other of the five points mentioned earlier, it is clear that the potential existing in national fisheries cannot be reached by private enterprise alone; a private entrepreneur is not in business to advance national goals such as the need to assure the food supply. This responsibility belongs to the government.

Indeed, since there is much justification for the assumption that a concerted government effort in the development of fisheries resources might be just as successful and rewarding as the effort involved in the development of agricultural resources, the ultimate question then arises; if potential benefits are so high, and the involvement of the government so indispensable, why has this involvement taken place so fitfully?

Administrative Reasons

Since seafood is one class of foods that can be used to satisfy man's hunger, one might ask why the responsibility for its assured supply did not come under the jurisdiction of the USDA as soon as this agency was established in 1862. The reasons appear obvious in retrospect. At that time, the emphasis was on the development of farms and farming, and there seemed to be no basis for connecting the production of seafood with farming activity. In addition, all seemed to be well with U.S. fisheries. At that time the United States was one of the world's leading fishing nations and seafood constituted one of its most important export items.

When the U.S. government established the United States Fisheries Commission in 1871, less than a decade after it had established the USDA, the charge given to the commissioner of fisheries was, first, to determine whether there was a basis for the allegations by fishermen that the fishery stocks were becoming depleted; and second, if the allegations were true, to make recommendations on how to resolve the problem.

Ever since then, the government has taken numerous actions to assist the U.S. fishing industry. For instance, the depletion of stocks—the reason for establishing the U.S. Fisheries Commission—was, in fact, determined to be a real problem (and it continues to be up to the present). Two solutions were, in time, proposed: One involved the operation of hatcheries and the other the imposition of quotas on fishermen working within the jurisdictional boundaries of the United States. But neither solution has been successful in addressing the problem. There may not be much more that can be done about it, however, because many of the factors that affect the biomass are beyond our control.

One such significant factor is the exploitation of the stocks by foreign fishermen in waters that lie outside the jurisdiction of the United States. In order to minimize this, in 1966 Congress extended U.S. jurisdiction from 3 miles to 12 miles of waters bordering its lands. When this action proved to be inadequate, the jurisdiction was extended to 200 miles in 1976. With this latest extension, a proportionately large share of the fish and shellfish resources of the world was brought within the control of the United States. This action will not, however, completely eliminate the effect of foreign fishing on U.S. marine stocks, because the stocks may and often do cross U.S. boundaries.

There are other factors that also affect the stocks but are beyond our control. Among these are a shortage of the foods required by the stocks for their survival, intensified exploitation by predator species, and adverse environmental effects. Some of these problems will remain; indeed, some can even be expected to worsen.

Other fields in which the U.S. government has, in some fashion, extended its assistance to the seafood industry are technology transfer, marketing, and even some minor forms of direct financial assistance.

The central point is that the U.S. government has taken numerous actions to assist the domestic seafood industry. But so far it has not taken either of two alternative paths that, if pursued consistently, almost certainly would help the United States attain the goal of assuring a satisfactory domestic seafood supply. One alternative is to insure that the fisheries agency—currently called the National Marine Fisheries Service (NMFS)—operate by the same philosophies that made the USDA so successful; the other is to delegate the mission directly to the USDA.

The first alternative amounts to a specific and direct formulation of a national policy. The second might be even more desirable for at least two reasons: first, because a successful philosophy is already in place and does not need to be developed anew; second, because when two agencies are engaged in similar activities, duplication of effort can overshadow potential benefits.

We have seen how the establishment of the USDA, the Land-Grant Colleges, and the State Experimental Research Stations, and the cooperative efforts among these agencies, the farmers, and other members of the U.S. agricultural industries resulted in phenomenal agricultural achievements. We have also seen that with the establishment of the United States Fisheries Commission, the United States did not make any progress toward assuring the U.S. seafood supply. Whereas seafood once comprised the bulk of U.S. exports, seafood imports currently contribute $5.6 billion to the U.S. trade deficit.

HOW TO OVERCOME THE LAG

From this brief review of the historical record it seems fair to conclude that the goal of an assured domestic supply of seafood cannot be achieved if we follow the present course. On the other hand, it is reasonable to assume that the potential can be reached either if, as suggested above, the NMFS adopts the philosophy of USDA or if somehow there is a merger between the two agencies. This proposed change would assure the United States its seafood supply just as the USDA did for the agricultural food supply.

If the nation should elect the first alternative, nothing else needs to be added to what has already been specified; with the second, however, a few more considerations deserve to be kept in mind. The merger would mean that the NMFS function should be moved from the United States Department of Commerce to the United States Department of Agriculture. Correspondingly, federal activities associated with the fresh waters' contribution to commercial fisheries should be moved from the United States Department of the Interior (USDI), where they are lodged now, to the USDA. The existing NMFS—and USDI—laboratories would either become extensions of, or otherwise be consolidated with, the USDA laboratories, thus making available to all laboratories the best of all three agencies in terms of specialized equipment and procedures as well as specialized scientific and engineering talent. The inspection service already operating out of the USDA could be expanded to include the inspection of seafood, since the protocols for inspection, sanitary codes, and so on, are the same in both cases. The USDA's excellent extension and public relations programs could be readily expanded to include a seafood aspect. Efforts in product development would also be facilitated, particularly product development involving a combination of products that are currently under the jurisdiction of either USDA or NMFS.

Our recommendation, then, that either the fisheries agency adopt the philosophy of USDA or that the two agencies be somehow merged into one follows naturally from the observation of the facts. While the differences between the two agencies are few and are mainly philosophical in nature, the similarities are many.

Similarities Between USDA and NMFS

The fish and shellfish that are harvested as seafood are animals, and in this respect NMFS functions are similar to those performed by the USDA for farm animals and poultry grown for food. They must therefore both be concerned with such disparate tasks as quality measurement and quality control; plant sanitation and sanitary procedures and equipment; food-borne diseases and their causes, identification, and prevention; and environmental toxins and the means to identify and to measure their concentrations. In addition, both agencies have an inspection function.

Both also have a utilization function and operate utilization laboratories to study and solve problems in preservation, processing methods, processing equipment, packaging, storage, transportation, handling quality control and quality assurance, the measurement and control of nutrition, microbiology, sanitation, environmental pollution, by-product utilization, waste control, and waste recovery, to name a few areas. Both agencies have a quality-grading function and must establish criteria that are based on similar grading principles.

They are both concerned with the marketing of food products and with the retailing of perishable foods as potential vectors of disease, especially because the different microflora of the products require avoidance of cross-contamination in retail outlets that perform some processing.

Both agencies have the opportunity of and may profit from working with those engaged in the production of the commodities over which they have jurisdiction: the USDA, with the farmers, the processors of agricultural products, the food retailers, the middlemen (wholesalers, jobbers, and so on), and those who supply the equipment to these constituents; the NMFS, with the fishermen, the seafood processors, the food retailers, the middlemen, and their respective suppliers of equipment and materials. Both agencies must deal with other federal agencies such as the Food and Drug Administration, the Environmental Protection Agency, the Department of Defense, the Occupational Health and Safety Administration, and other federal, state, and municipal agencies in much the same way and for many of the same reasons. Both have the opportunity of and may profit from collaborating with universities: the USDA with the Land-Grant Colleges; and the NMFS, with the Sea-Grant Colleges.

REFERENCES

Power, E. A., and W. L. Peck. 1971. The national picture. In *Our Changing Fisheries*. Shapiro S. (Editor). Washington, DC: Superintent of Documents.

Rhodes, F. H. T. 1985. To gain a market edge. *The New York Times*, November 2, p. 25.

Stansby, M. E. 1971. Fishery food science. In *Our Changing Fisheries*. Shapiro S. (Editor). Washington, DC: Superintendent of Documents.

Stefferud, A. (Editor). 1962. *After a hundred years: The Yearbook of Agriculture*. Washington, DC: Superintendent of Documents.

Tannahill, R. 1974. *Food in History*. New York: Stein and Day.

U.S. Department of Commerce. 1975. *Historical Statistics of the United States, Colonial Times to 1970. Part 1*. Washington, DC: Bureau of the Census, U.S. Department of Commerce.

ASSURANCE OF
SEAFOOD QUALITY

4

Characteristics of Seafood Quality

DEFINITION OF TERMS

A number of terms used in the seafood industry may be strange to some readers; others may have meanings that are dissimilar to those in common usage. In order to eliminate any ambiguity we will define the specific meaning of these terms.

The term *seafood quality* is generally interpreted to mean the quality of seafood, without reference to the level of quality. Actually, the term has a more specific meaning; it refers to the eating quality of seafood. *Quality seafood*, on the other hand, is generally interpreted to mean a seafood of high quality. Again, the meaning refers to eating quality. The technical term for eating quality is *organoleptic quality*, because quality is perceived through the use of the sense organs (eyes, nose, taste buds, and tactile elements); another term with the same meaning and used frequently is *sensory quality*. Throughout this text, seafood quality means eating or organoleptic quality.[1]

---

[1] The emphasis on organoleptic quality does not stem from a lack of concern for such other areas as the health and safety aspects of seafood. Seafood safety has not been found to be a problem in any of the surveys of quality that are cited in chapter 7; and the analysis and measurement of the safety of seafood, although often complex to perform, are relatively reliable. On the other hand, all of the known methods of analysis and measurement of the organoleptic quality of seafood suffer from varying degrees of unreliability (see chapter 5). In addition, as evidenced in the surveys cited in chapter 7, it is the organoleptic quality of seafood that has been criticized.

The term *fresh seafood* may be ambiguous, because in the trade parlance it refers to seafood that has never been frozen. Accordingly, even seafood in varying stages of spoilage may be called fresh seafood, provided that it has never been frozen. Since the word *fresh* is defined as recent or not spoiled, one can readily understand the confusion that may result when partially spoiled fish is referred to as "fresh fish."

The expression *slacked-out seafood* refers to seafood that is not in the frozen state but had once been frozen then thawed and thereafter handled as fresh seafood. Although the products may appear as fresh fish once they have been thawed, legally they cannot be labeled as fresh.

Frozen seafood is used to categorize those products in which sufficient water molecules have been frozen; consequently, the products have acquired a rigid structure that resists deformation under heavy hand pressure.

The phrase *superchilled seafood* is unique. It is used to describe seafood that is too cold to be accurately described as fresh and not cold enough to be called frozen. Other terms that have been used interchangeably with superchilling are *supercooling, light freezing,* and *partial freezing.* Seafood that is held in the range of temperatures where superchilling occurs ($-3°$ C to $-1°$ C or 26.6° F to 30.2° F) will not, as a rule, be frozen to such a degree that it is rigid at any time during its distribution. For this reason, although it is a controversial practice, superchilled seafood (which is generally not marketed under this name) is eventually sold and properly labeled as fresh.[2] In the range of superchill temperatures, some water within the product will freeze, lending a degree of structural rigidity to it, but not enough ice will be formed to prevent deformation when the product is subjected to heavy hand pressure. Those readers who might have an interest in how decreasing the temperature affects the physical state of seafood should turn to appendix 2.

WHAT IS KNOWN ABOUT QUALITY?

It is safe to say that not many people are familiar with the criteria that are used to determine the quality of seafood. The reports from relevant surveys have indicated that the population living along a coast and adjacent regions comprises the bulk of those who have enjoyed

[2] It is noteworthy that superchilling has been used to preserve the quality of fish at sea over relatively long periods, although there is some disagreement as to the practical effectiveness of this process. The results of some experiments have shown that the deteriorative effects of the superchilling process on the quality of seafood exceed its preservative effects. There is a physical reason for the potentially harmful effects of superchilling on quality: The damage incurred by tissues during the freezing and thawing cycles and the frequency of these cycles due to fluctuations in temperature cannot be avoided.

high-quality seafood. With the proliferation of national fast-food franchise restaurants, which have established a reputation for serving high-quality seafood sandwiches, and with the diffusion of restaurants serving high-quality seafood entrees, however, this geographic limitation has already started to disappear.

It should be noted that, except for conventional restaurants, fast-food chains (and in the view of the authors, the McDonald's chain especially) in just a few years have done more than any other element in our society to enhance the image of seafood. Furthermore, the very high quality of the fish used in fast-food chain sandwiches has completely dispelled the long-standing false notion that frozen fish is not of as good eating quality as fresh fish. These chains have achieved an unusual level of success in merchandising fish. The key to their success is that they control the entire distribution chain, from the buying of the highest-quality fillet blocks to the preparation step just before consumption. In fact they have achieved exactly what we are prescribing for the rest of the seafood industry. They assure quality where it counts—at the point of consumption! And, just as important, they do it with consistency.

CRITERIA OF QUALITY

When foods are consumed, their quality is perceived through the conscious or subconscious integration of their sensory or organoleptic characteristics. These characteristics may be grouped under the headings of appearance, odor, flavor, and texture (at times odor and flavor are combined into one).

The quality of all seafood is judged by the same criteria. There are two issues with which one must be concerned: the process of judging—or more specifically, the measurement of quality—and the criteria for judging quality. The measurement of quality will be discussed later; in this chapter, the discussion is limited to the criteria used in the evaluation of the quality characteristics of fish fillets.

However, since fish fillets are cut from whole or eviscerated fish, and since the quality of the fillets is dependent on the quality of the fish, it is also relevant to call attention to the criteria upon which the quality of whole fish is evaluated. As noted above, the quality characteristics are grouped into four classes, and we shall accordingly discuss them by class and in the order that they are usually encountered during the process of organoleptic evaluation.

Appearance

In most cases, the first opportunity to evaluate the quality of a seafood is governed by its appearance. This is true whether we see the product through the glass of a display counter or in a clear plastic wrapper, or cooked as an appetizer or an entree. We will discuss the appearance of whole fish first and then that of fish fillets in the respective market forms in which they most commonly appear.

Whole Fish. The market forms in which whole fish are mostly distributed to the consumer are as follows: the larger ones, about 2 kilograms or 4.4 pounds, such as cod and some mackerels, are usually sold as fresh (unfrozen) or slacked-out (once frozen but now unfrozen); the smaller ones, usually less than 1 kilogram or 2.2 pounds, such as some mackerels and even smaller ones such as smelts, may also be sold as frozen. Some fresh whole fish are simply over-wrapped or placed in plastic pouches. Those in clear wraps may still be judged for appearance, although not to the same degree as unwrapped fish.

In the evaluation of fresh whole fish, the experienced individual recognizes each by species (specific coloring, shape, and other physical attributes). In high-quality whole fish, the skin has a noticeable sheen, which is a remnant coating of clear slime that covers the fish in life. The eye fluid is clear, and the eyes are neither discolored (no blood spots, and so on) nor deformed. The gills are from a deep to a bright red color. In wrapped fish, one should look for free fluid within the package— and consider both the amount of the fluid and its appearance. In high-quality fish there should be little or no fluid, which in any case should be clear and with little or no color. When packaged whole fish is of less than good quality, the package will contain a considerable amount of fluid. It is usually cloudy and noticeably colored, often with blood. Some packages may contain an absorbent layer. In such packages, when the fish is of high quality, the absorbent layer will appear relatively dry and not discolored; with fish of less than high quality, the absorbent layer will show the presence of noticeable amounts of absorbed fluid, which may also be discolored. When packages containing an absorbent layer also contain a relatively large amount of cloudy, colored fluid, the product is definitely not of high quality.

The appearance of whole frozen fish that are either over-wrapped or in pouches made of transparent plastic is sometimes difficult to assess when there is an accumulation of frost inside the package. Excessive frost means excessive water loss by the fish. In high-quality fish, there is little or no frost, and there is no discoloration of the product.

Fish Fillets. Although most fillets are marketed as fresh or frozen, they can also be found in the form of frozen seafood entrees. Fresh fillets may be displayed either on ice in single layers or as small piles of fillets resting on an ice layer; as a result of experiments showing the advantages of tray packing, they may also be marketed in plastic over-wrapped trays containing an absorbent layer.

As can be seen from figure 4.1, fillets have a distinctive shape. It should be noted that fillets cut from flat-bodied fish are much broader than those represented in this figure. The outline of the fillet should be clean-cut and not appear ragged. There should be no separation of the tissue, and the surface should have a glossy appearance. The color should be typical the species (e.g., Atlantic cod should be white), and there should be no discoloration or blood spots. Of course, experienced individuals will know which fillet color to expect for each species, whereas inexperienced buyers may have some difficulty with most species.

The discussion on the presence of fluid and frost in packaged whole fish applies to fillets as well. The typical fillet shape should also be apparent when fillets are cooked. Because of the size of portions ordinarily served by restaurants, however, larger fillets are usually cut transversely, and the shape is not relevant in evaluating appearance in such cases. The color of most cooked fillets should be white, but there is a tinge of gray in some species such as pollock and gray sole and a tinge of pink in others, such as ocean perch.

When fillets are cooked, they are easily *flaked*, that is, the muscle segments (the "flakes," technically called *myomeres*) can be separated with a fork because of the instability of their connective tissue when exposed to heat. When the flakes are parted, they should have a glossy appearance with little or no particles of curd between them, because the curd is indicative of the progress of deteriorative reactions.

In some cases, it is difficult even for experts to identify the species of fish fillets. It would be nearly impossible to distinguish between fillets of equal size from Atlantic cod and haddock, for example, were it not

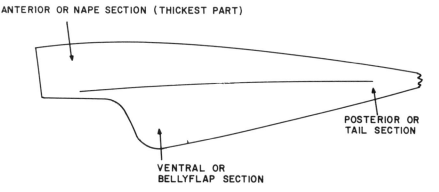

Figure 4.1. Typical Shape of Fillet Cut from a Round-Bodied Fish.

for the fact that cod fillets are usually skinned (the skin is removed), whereas the skin on haddock fillets is usually left on.

To make things more complicated, the legal nomenclature of seafood, established by the Food and Drug Administration (FDA), is either ignored or circumvented in some cases. For example, Atlantic pollock has been merchandised as Boston bluefish and what is sold as canned sardines often is actually canned herring. One relatively new term that has remained a source of confusion to consumers is *scrod* or *schrod* (neither is a species of fish).

Seafood processors in New England generally use scrod to refer to the fillets of young haddock (about 1 kilogram or 2.2 pounds in weight). However, they also call scrod those fillets that come from similar-sized cod and pollock—and identify them as cod scrod and pollock scrod. There appears to be no confusion with the use of the term by either processors or fishermen, even though their particular definition is not in the dictionary. Seafood retailers and restaurateurs, however, at times use scrod to refer to fillets or pieces of fillets from cod or haddock of any size without any indication as to the species. In fact, there is reason to believe that a significant number of consumers are under the assumption that scrod is a species name, like cod or haddock.

In one marketing experiment, haddock fillets were sold more readily when they were labeled scrod (to repeat, a relatively new term) than when they were labeled haddock. A result such as this tends to support the contention that scrod is confused with the name of a species by some consumers, especially since all, or nearly all, restaurants are apt to have scrod, haddock, and cod entrees on the same menu. Since restaurants usually serve only high-quality seafood, it can be understood why scrod eaten in restaurants leaves a better impression than either haddock or cod, species that, especially if bought in a supermarket, are most likely associated with negative experiences of the past. One New England–based supermarket chain has taken a constructive step in the use of these terms. They have advised the public that when they advertise *scrod* fillets, the fillets have been cut from cod and when they advertise *schrod* fillets, the fillets have been cut from haddock, with the "h" in schrod representing the "h" in haddock.

Odor

The odor of freshly caught fish is mild. It has been described as typical of the sea and seaweeds. If fish are held in ice from the time they are caught (assuming they were caught by acceptable methods, a subject to be discussed later), they will remain of relatively high quality for a period of about one week. During all this time, there will be little or no objectionable, or "fishy," odor. One of the causes of the fishy odor is that it is detected when the trimethylamine oxide in fish tissue is reduced to trimethylamine, usually by bacteria described as facultative aerobes. These are bacteria that, even though they do not necessarily require oxygen, during scarcities of atmospheric oxygen are able to satisfy their requirements by scavenging it from compounds to which oxygen is bound. Other objectionable odors are due to the bacterial production of ammonia, sulfides, and so on.

The odor of seafood should be judged only when the products are in the unfrozen state. In the frozen state, the odor is not completely detectable, and the fish odor that is evident in some of the packaged frozen products may come from the outside of the package. Therefore, if the products are frozen they should be thawed and, if packaged, they should be removed from the package. Only then can an evaluation of quality, using odor as a criterion, be made. When of high quality, the thawed products should have either little or no odor.

Flavor

Because fish and fish fillets sold in supermarkets are not generally consumed raw in the United States, the flavor characteristics of these products in the raw state are not relevant to our discussion. Therefore we will be concerned only with the flavor of fish and fish fillets in the cooked state. Cooked fish and fish fillets having a low fat content (less than 1 percent), such as cod, haddock, cusk, hake, and most flat fish, have little or no flavor when they are of high quality. There should be no off-flavors (rancidity) nor acid or bitter taste, however mild because this would indicate that the quality has deteriorated to some degree. On the contrary, in most cases when their quality is high, seafood has a mildly sweet taste due to the presence of a small amount of sugar in the flesh. The quantity of sugar that remains depends mainly on the amount of struggling done by the fish when it was caught: the greater the struggle, the less the amount of sugar that remains. The sugar, even if present, is broken down during storage, and eventually none will be detected.

Fish and fish fillets having a high fat contents (6 percent or more), such as mackerel, herring, salmon, and sardines, are more flavorful than lean fish. The flavors of these products are pleasant as well as unique, but only while their quality is high. As tasty as the fatty fish are, however, their fats oxidize rapidly and produce flavors, described as rancid, that are objectionable to most people. (The rate of oxidation is directly proportional to the temperature, to the chemical structure of fats, and to the availability of oxygen.) Experienced tasters can identify rancid flavors even before they accumulate to amounts that are objectionable to the average consumer.

Since it does not require much oxidized fat to manifest rancidity, even fish containing low or intermediate amounts of fat (2 to 5 percent) soon manifest rancidity. Fish that are relatively lean (less than two percent) but have concentrations of fats, such as the subdermal layer of fat in swordfish and halibut, also manifest rancidity. In some cases, the rate of formation of rancidity is so fast that the taste of the fish, an exquisite taste when the fish is just caught, becomes intolerable within a few hours. Such is the case with the red meat of tuna. When the meat is consumed soon after the fish is slaughtered, its flavor is so delicate that it has been compared to that of filet mignon. Yet, within a few short hours, the nature of the rancidity that is produced is such that the product becomes intolerable to most palates. Another case is the menhaden, used largely for the production of fish meal and not considered as food fish. Yet, when prepared just after catching, it is reported to be among the best tasting of all species.

Clearly, the rancid flavors that develop in the different species have different effects on the organoleptic acceptability of the products. When rancidity occurs, the effect appears to be more adverse in the fatty species than the lean ones. A product generally loses acceptability with the appearance and progression of rancidity, but not for everyone. Having little or no familiarity with the taste of fresh fish, some people prefer the taste of rancid seafood. Rancidity occurs even during frozen storage, although the rate is slower at lower temperatures.

Texture

The final criterion used in the organoleptic evaluation of fish, as well as other foods, is texture. Texture is related to the physical properties that are experienced during biting and chewing. Although this criterion is more relevant when it is applied to cooked fish, texture tests are made routinely by inspectors on raw fish because they are indicative of the texture of cooked seafood.[3]

An easily evaluated property of the texture of raw seafood is the resiliency of the tissue. In high-quality fish, when the tissue is depressed, it resumes its original shape as soon as the pressure is removed. In fish of poor quality, most of the depression will remain long after the pressure is taken off.

The texture of cooked fish is most appropriately determined during its consumption. It should be relatively tender, not tough, a characteristic

---

[3] Texture during and prior to rigor mortis is usually of no relevance to the consumer.

that may occur in fish held in frozen storage under adverse conditions or may derive from overcooking. It should not be dry, which may occur in fish held in frozen storage when either the holding temperature or the package is inadequate or, again, may derive from overcooking. It should not have too soft a texture (often described as mushy), nor should it have a slimy feel: Both characteristics are indications of enzymatic or microbial deterioration.

REFERENCES

Carlson, C. J. 1969. Superchilling fish: A review. In *Freezing and Irradiation of Fish*. R. Kreuzer (Editor). London, England: Fishing News (Books) Limited.

Gould, W. A. 1977. *Food Quality Assurance*. Westport, CT: AVI Publishing Co.

Kramer, A., and B. A. Twigg. 1970. *Quality Control for the Food Industry, Vol. 1, Fundamentals*, 3rd ed. Westport, CT: AVI Publishing Co.

Stansby, M. E. 1963. *Industrial Fisheries Technology*. Huntington, NY: Robert E. Krieger Publishing Co., Inc.

Thorner, M. E., and P. B. Manning. 1983. *Quality Control in Food Service*, revised edition. Westport, CT: AVI Publishing Co.

5

Measurement of Seafood Quality

Under current practices, the sensory quality of seafood sold in super-markets is measured largely by the processor, and once the product leaves the processing plant, little or no additional inspection takes place. The sensory quality is that which is sensed by sight, smell, taste, and feel. Since seafood is not subject to mandatory inspection, it is possible that quite a bit of that sold through supermarkets may not be inspected at all. Even if it is inspected by the retailer, however, current practices cannot be considered adequate for assuring its sensory quality.

In practically all cases, the analysis for sensory quality is performed by a single individual using a two-point scale to determine whether the product is acceptable (1) or not acceptable (2). This has been, and still is, the standard policy.

Striving for accuracy in the sensory analyses of the quality of seafood, scientific laboratories also use objective (physical or chemical) tests. Sometimes they find it necessary to employ sensory analyses using groups of individuals rather than a single person. These groups are called *taste panels*, although, in fact, the analyses involve more than just the evaluation of the taste of the product. The reason laboratory analysts employ more sophisticated analytical methods is that they need to know not only whether the samples under examination are acceptable, but also the level of acceptability and the reliability of the measurement.

SENSORY AND OBJECTIVE TESTS FOR MEASURING SEAFOOD QUALITY

The tests used for the measurement of seafood quality may be divided into two categories: One encompasses the sensory tests; the other, the objective ones.

Sensory tests are employed to determine sensory quality. They depend on the observer's ability to accurately assess the appearance, odor, taste, and texture of a product. A certain amount of personal opinion may influence the interpretation of results. The evaluation of the quality of seafood sold in supermarkets is accomplished mainly by sensory tests. The results of objective tests, most of which employ physical and/or chemical principles, are not influenced by the observer. It is this aspect of the tests that lends credibility to their results and makes them so important in research as well as in legal proceedings. To obtain a better understanding of both test categories, we shall compare the sensory with the objective from several viewpoints.

Reliability of the Tests

The reliability of any test depends on the accuracy of the measurements and the precision of the method used for testing. Those readers who are interested in the mathematical elements of reliability are referred to appendix 3.

SENSORY TESTS

Sensory tests may be performed using large consumer-type panels; small expert panels; small experienced panels; single expert individuals; and single experienced individuals.

Large Consumer-Type Panels. In most cases, the quality of seafood is measured to determine how it will be judged by consumers. The most accurate way to make this measurement would be to have a sample tested by all consumers, omitting no one. If the consumers were then asked to rate the quality of the product on a scale of 1 to 5 or 1 to 10, and if all of the scores thus obtained were plotted to show the number of people assigning each of the scores, we would obtain a relatively symmetrical, unimodal (one-hump) curve with a bell-like shape (see figure 5.1). This is called the normal distribution curve, and it tells us a number of things, the most important of which is that not all consumers react the same way in any test. Even those products liked by most of the population may be disliked by some. The curve tells us that there is no way to assure that a product will be well received by the

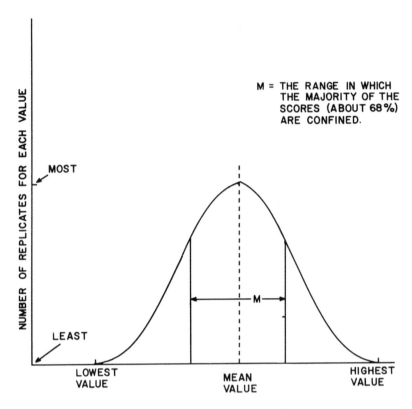

Figure 5.1. The Normal Distribution Curve.

entire population. But it also tells us how the majority of the population rates the product, because the upper portion of the hump on the curve expresses the values of the majority decisions (about 68 percent of the total population). Since one cannot ever please everyone, the next best objective is to try to please the bulk of the population. Thus when one measures the quality of a seafood product, no matter which test is used, one tries to determine how the majority of the population will judge it.

Obviously, we cannot hope to get the response of the total population on the quality of a product, so we have to turn to other means. We can use a sample of the population, but unless the sample is sufficiently large, we should not expect the same curve as in figure 5.1. The curve is likely to be skewed (not as symmetrical), the hump may not be as well defined, and, most important, the bulk of the sample response may be displaced to one side of the response that would be given by the total population. If different samples of the total population, each different in number, were to be asked to rate a product, the largest sample would yield a response most nearly resembling the response of the total population. For this reason, consumer-type panels usually number in the hundreds to thousands, depending on the size of the population for which the product is to be targeted. Panels of such size usually produce a reliable measure of the total population, and can also be considered to be a relatively accurate tool for measuring the organoleptic quality of seafood.

Small Expert Panels. A small expert panel consists of as many as twelve individuals who have received training in sensory testing. Scientists need to measure the quality of seafood for a number of reasons. For example, they need to know how the quality of seafood is affected by various preservation, processing, packaging, and storage techniques. Although much of the testing of quality is done by chemical or physical (objective) tests, at times, small expert panels are also employed to gather corroborative sensory data to support that obtained with other tests. Nearly all objective tests are evaluated by comparing them with sensory data. When these tests correlate well, they are considered to be reliable. Thus scientists who either develop objective tests or are interested in calibrating an objective test must rely on sensory data. Since most scientists have neither the resources nor the time to employ large consumer-type panels, they use a small expert panel from within their facility. In order to compensate for the small size of the panel, these panelists are trained to the extent that they are considered to be experts.

Small Experienced Panels. A small experienced panel consists of as many as twelve individuals who are experienced in sensory testing but who are not experts. In some cases, the research scientists who use small sensory panels do not train the panelists adequately. However, the panelists do gain experience with time, and although they may not perform as well as experts, they are better qualified to judge the quality of seafood than are inexperienced individuals. Yet because of the small size of most of these panels, and because the panelists are not expert enough to compensate for their small number, this type of panel usually fails to determine quality with adequate precision, causing the results to be in conflict with the results of objective tests or with other known facts. Consequently, it is this type of panel, from which too much is usually expected, that has given sensory testing its notorious reputation for unreliability.

Single Expert Individuals. An expert meets two criteria. He or she is well-trained and has been approved by an appropriate authority to practice as an expert. These criteria are met by federal inspectors as well as those trained and approved by the government. Theoretically, the inspectors use a four-point scale (see table 5.1) corresponding to

Table 5.1   Sensory Standards Used by USDC Inspectors

Grade*	Criteria
A	Good flavor and odor characteristics of the species
B	Reasonably good flavor and odor characteristics of the species
C	Minimal acceptable flavor and odor characteristics of the species
Substandard	Meets either the grade A or B criteria in sensory tests, but not in non-sensory tests. The non-sensory criteria include meeting the declared weight, freedom from defects, and correctness of nomenclature.

*Products that have objectionable flavors and odors are not eligible for grading.
Source: Wildlife and Fisheries no. 50, 1984.

the four grades of quality: A, B, C, and substandard. There are specific sensory and non-sensory criteria for the first three grades. Any product that cannot meet the non-sensory criteria referred to in table 5.1 cannot be graded A, B, or C and that accounts for the fourth point on the scale (substandard). Even substandard products must meet the minimum sensory criteria. Expert sensory testers are generally adequate for the job at hand, but cannot be considered to have adequate precision for scientific testing.

Single Experienced Individuals. Single experienced individuals who evaluate the sensory quality of seafood include such members of the industry as buyers, brokers, and processors. This type of sensory analyst works with a two-point scale: acceptable or unacceptable. Different individuals have different standards as to the criteria for acceptability, therefore there could easily be differences of opinion between any of them. Still, these individuals are quite experienced, and in fact, some are so capable that they could, if necessary, work with scales having more than two points.

Whereas most individuals who evaluate the sensory quality of seafood without collaboration (be they inspectors or buyers) fulfill their responsibilities reasonably well, single judges are never used in a research laboratory. The main reason is that multiple scores obtained from panels tend to compensate for deficiencies in the human response. Single scores do not offer this opportunity.

OBJECTIVE TESTS

Objective tests usually employ instruments or devices whose results are not influenced by the observer. An example of this is the use of a scale to obtain the weight of an object (if one were to estimate the weight simply by hefting the same object, it would be a subjective analysis). Most of the objective tests employ chemical or physical principles, and therefore are considered to be reliable and have the most desirable degree of credibility, especially in legal proceedings. It is mainly because of their reliability that research scientists prefer to employ objective tests whenever possible. Because of the importance of these tests, those that have been accepted as reliable through a nationwide program operated by the Association of Official Analytical Chemists (AOAC) are listed in that organization's manual which is revised annually.

Sensory Versus Objective Tests

There are three aspects of reliability that have yet to be discussed. The first is that sensory panelists are subject to a fatigue phenomenon; that is, their judgement is affected when they attempt to analyze more than a saturation level of samples (about six samples) at one sitting. Objective methods are not affected by the number of samples to be analyzed. The second is that sensory results are not always reproducible, whereas the results obtained with objective tests always are. The third is that most objective tests gain acceptance only when they correlate well with sensory analyses. This characteristic should raise a few questions about the dependency on objective tests and the widespread conviction of the unreliability of subjective tests.

Cost of Tests

The cost of conducting sensory tests is variable. When large consumer-type panels are used, each test may cost hundreds to thousands of dollars because of the number of man-hours involved. With small expert panels, on the other hand, the cost is much lower (usually less than $100.00), even though the initial costs of training the panel and of maintaining its acuity might be considerable. The cost for small experienced panels may be about the same as for small expert panels, although the training costs for the former may be lower. Single individual costs are nearly negligible. In a government/industry study (Gorga et al 1978), inspection costs were put on a weight basis and were found to be $.006 per pound.

The costs of objective tests are also variable, but most are relatively inexpensive. For example, routine chemical tests for trimethylamine, ammonia, volatile acids, and so on, can be carried out by a chemist in hours, and analyses to measure color or texture may take only minutes and cost relatively little. On the other hand, an analysis by computer-assisted chromatography and mass spectrometry is complex and may cost a few hundred dollars.

Flexibility of Tests

A sensory test yields a result that integrates all of the quality attributes of seafood (appearance, odor, flavor, texture). Most objective tests, however, cannot perform this integration because they are designed to yield information on the quality of only one attribute. For example, a colorimetric test can help to evaluate only appearance, a tenderometric test to evaluate the texture; many of the chemical tests are designed to measure either one compound, such as ammonia, or a class of compounds such as carbonyls, sulfides, and volatile bases. A few objective tests are reported to measure the overall sensory quality (appearance, odor, taste, and texture) of the product. The electronic fish tester is one example of an objective test that is reputed to have this capacity.

Sensory analysis can be used to measure the quality of seafood at any stage of its shelf life (with some limitations), but most objective tests are not useful until the spoilage compounds, which they are designed to measure, become evident. By this time, of course, the quality has reached varying levels of deterioration.

Relevance of Tests

If the purpose of testing the quality of a seafood is to determine how the product will be received by consumers, then sensory testing is more relevant than objective testing. In sensory testing, quality is perceived by the same systems that are used by consumers—the human senses. Objective tests are only indirectly relevant, since their systems depend on a mathematical correlation with sensory data in order to demonstrate that they can be used to measure sensory quality.

If the purpose of testing, however, is to follow the development of a particular change (e.g., change in color, accumulation, or dissipation of a chemical compound, the effect of a treatment on a particular aspect of quality), then objective tests, because of their greater accuracy in measurements of this type, are more relevant than sensory tests.

Applicability of Tests

When the purpose of measuring seafood quality is to determine its level of acceptance by consumers, the most accurate test is the large consumer-type panel. But because of its high cost, such a test has only limited applicability. As stated earlier, industry usually uses only one individual to measure quality by sensory analysis, although in some cases, objective tests are also used.

In research laboratories, the tests having the widest application are objective tests and sensory testing with small panels. From the scientific point of view, the objective tests are quantitative and more accurate than sensory tests, but sensory tests are necessary because they are the standard with which objective tests must be correlated.

Justification for Continued Use of Sensory Testing

Although continued technological advances are constantly employed to develop more and better objective tests, the use of sensory analysis for the measurement of seafood quality remains very important. It is the method used most often by industry because it is so inexpensive and effective, since it depends only on a two-point scale: acceptable quality and unacceptable quality.

From the standpoint of research laboratories, sensory testing is important because it is used to corroborate the data obtained with objective tests and as a standard with which to evaluate the reliability of these tests. However, scientists expect, and too often assume, a greater precision and accuracy from the small sensory panels than these can yield. This is the source of much of the disappointment in, and mistrust of, sensory data.

When it is necessary to determine the acceptable shelf life of seafood in order to determine how long it can be safely kept free from a particular health risk, the most reliable method to use is sensory analysis, with either a large consumer-type panel or a small expert panel. Neither small inexpert panels nor individual inspectors, no matter how expert, should be used in this case (Ronsivalli et al 1971).

Since sensory testing methods are basic in the measurement of seafood quality, it is important to explore their few inherent deficiencies, which account for most of their notorious unreliability. In addition, it is reasonable to consider at least one proposed method for improving the accuracy and precision of small sensory panels, because they are especially vulnerable to unreliability.

DEFICIENCIES INHERENT IN SMALL SENSORY PANELS

Inherent in sensory testing are three major deficiencies that account for its reputed unreliability: variability in the human response, lack of information on reliability of small sensory panels, and use of faulty score sheets.

Variability in the Human Response

Each individual perceives the quality of seafood (and other food as well) differently. This is due to at least three reasons. The first is subjectivity, which reflects the integrated sum of personal opinions, personal preferences, customs, experiences, etc. The second reason is objectivity, which reflects the keenness of the senses and the ability to identify what is sensed. Some have a keen sense of smell, others do not; some have good eyesight and can differentiate among colors, others have poor eyesight or are color-blind; some have a keen sense for tastes and flavors, others do not. The third has to do with that periodic quirk in the human performance, described as an "off-day," that accounts for those times when performance dips below personal par.

It is because of these variables in human response that panelists are apt to use most, if not all, of the values found in the score sheet. Consequently, the data from sensory tests involving a large consumer-type panel yield bell-shaped plots similar to the one shown in figure 5.1. It is understandable that scores from such tests cover a broad range of values. It is also understandable that small panels will probably not yield results as accurate as those generated by the large panel.

Lack of Information on Reliability of Small Sensory Panels

The reliability of any device used in making measurements depends on its accuracy and precision (see appendix 3). In too many cases, sensory data generated in scientific investigations are subjected to statistical analyses to demonstrate correlations with other data, such as those that might result from objective tests; but whereas the accuracy, precision, and reproducibility of the objective test methods are relatively simple to determine, the same usually cannot be said for the sensory panels. It was pointed out earlier that the reliability of most, if not all, of the objective tests used to measure the sensory quality of seafood is invariably proven by showing a correlation between the results of both the objective and the sensory tests.

Use of Faulty Score Sheets

Although it may appear that the omission of information on the reliability of the small sensory panels used in scientific investigations is an unforgivable oversight, an analysis of the score sheets used by these panels will quickly show the basic reason for it. Most, if not all, of these score sheets rely on the use of unquantifiable and ambiguous terms to define too large a number of levels of quality. Table 5.2 contains a verbal

Table 5.2. A Typical Score Sheet Used by Small
Panels

Level of Quality	Numerical Value
Excellent	9
Very good	8
Good	7
Fair	6
Borderline	5
Slightly poor	4
Poor	3
Very poor	2
Inedible	1

list used by panelists to rate the quality of a seafood (the numerical list is used by the analyst to analyze data mathematically). It can readily be seen that the terms in the verbal listing are ambiguous. For example, what is "good" quality? To some "good" means the opposite of "bad." Therefore, a quality that is "not bad" must be "good," or, in this case, it can be "fair" or "borderline." But what is "borderline"? These terms are also dependent on personal interpretations. "Good" quality might be interpreted as "reasonably good" and vice versa, and "reasonably good" might be interpreted as "minimally acceptable" and vice versa.

A second problem associated with the verbal terms in the score sheets in table 5.2 is that they are not quantitative expressions and therefore cannot be used in any analysis to establish either the accuracy or the precision of the panel. Nor can they be used in the analysis of the data that would permit a determination of the quality of the product tested. It is accepted practice to resolve this problem by assigning numerical values to the verbal terms, and using the numerical values, such as those listed in table 5.2, in the subsequent analyses of the data. But this is not good scientific practice, because the ambiguity of the verbal terms is passed on to the corresponding numerical values and will remain and affect any subsequent analysis involving these values. Another possible problem with the numbers is that they are assigned in a whole number sequence, as though each quality level were exactly one unit greater or less than its neighbor.

Finally, there are too many terms listed, and even if not ambiguous, their number implies a precision that has not been determined to exist for sensory testing. The only reason that quality testing in the industry is reasonably reliable is that the evaluation is done with a two-point scale, acceptable or not acceptable. Even federal inspectors, the officially authorized experts, do not have to deal with a large range of levels of quality since their scale has only four points (Grades A, B, and C, and Substandard). They too may sometimes unconsciously resort to a two-point scale in their sensory examination since other, more quantifiable criteria may be the actual determinants of the grade of quality.

It is, therefore, in the sensory analysis in scientific investigations that there arise doubts about the reliability of sensory testing. Too much is expected from these panels relative to their actual capabilities, and too little is known about the actual precision and accuracy of each one of these panels. It is like measuring the dimensions of this page with a standard ruler that is divided into centimeters, with each centimeter subdivided into millimeters, and expecting to obtain values to the nearest micron.

Because of the importance of sensory testing in determining the quality of seafood and the need to refine it to improve its potential contribution in research and in the development and/or evaluation of objective tests, we recommend continued efforts to improve its reliability. It is obvious from the foregoing that the area of sensory testing requiring the greatest effort is that of small panels in research. But it should be kept in mind that improvements in that area will benefit other areas of sensory testing as well.

IMPROVING THE RELIABILITY OF SMALL SENSORY PANELS

What follows is a description of one attempt to improve the reliability of small sensory panels. First of all, it was obvious that the score sheet used by the panel had to be redesigned so that quantifiable terms would define organoleptic quality. Second, it was also obvious that small sensory panels require specialized training and that it was necessary to evaluate their reliability. Thus, a two-part effort was initiated, the first to develop a quantifiable score sheet, and the second to train and measure the accuracy and precision of the panel (Learson and Ronsivalli 1969).

The Redesigned Score Sheet

Because it had already been established that the rate of the deterioration of quality in fish is directly dependent on the storage temperature, it became a simple matter to consider increments of time as descriptors of the level of quality. Since fish have a shelf life of about two weeks at 0° C (32° F), and since the recommended storage for fresh seafood is 0° C (32° F) or as cold as possible without freezing, the proposed score sheet was based on the approximately fourteen days that it takes fish to spoil at that temperature. Thus, with a time-based score sheet panelists did not have to describe the quality of a seafood in arbitrary and ambiguous terms; they would only have to estimate for how many days, after death, it had been stored at 0° C (32° F). To make this evaluation, of course, they would have to recognize the daily changes that occur in fish. Quite likely it would be necessary to observe these changes for each species that might be tested because of species differences that exist for fish and shellfish, although such differences should not be expected to be great. A scale showing this relationship is shown in figure 5.2. It can be quickly seen that the sensory values used by the panelists to rate the quality correspond exactly to the number of days in storage at 0°C (32° F). The terms are quantifiable and unambiguous, and the changes occurring daily are accurately defined by the numerical sequence assigned.

Development of an Expert Panel

Using a scale similar to the one shown in figure 5.2, the National Marine Fisheries Services (NMFS) Laboratory in Gloucester, Massachusetts developed, and tested for its reliability, a small expert panel (Learson and Ronsivalli 1969). Training of the panel was done in two phases. In the first, as many prospective panelists as could be brought into the exercise were allowed to observe cod fillets that ranged in qual-

Days of Storage
at 0°C (32°F)

1
2
3
4
5
6
7
8
9
10
11
12
13
14

Figure 5.2. A Quantitative Score Sheet.

ity, in increments of one day, from the freshest possible (about one day after slaughter of the fish) to those that were definitely spoiled (held at 0° C or 32° F for about two weeks after slaughter of the fish). The examination depended on the appearance and odor of only the raw fillets. Each day the prospective panelists observed fillets at varying storage times, with elapsed times explicitly stated. The purpose was for the trainees to use whatever descriptive terms best suited them, provided these terms would help them identify the changes observed and place these changes in relation with the times of storage. In other words the panelists made up their own individual lists of descriptive terms upon which they could depend to determine the length of storage of the fillets. (It is true that the panelists were entrusted to convert a subjective integration of quality into a quantifiable term, but in this instance they had the advantage of using the terms that were most helpful to them, and their ability to make reliable determinations could be objectively tested.)

During that initial phase, the prospective panelists were able to check the reliability of their respective lists of descriptive terms and to make alterations as necessary. After a few weeks, most panelists had reached a familiarity with the new rating technique, and the second phase of the training was started. In this phase, everything done in the first phase was continued, except that some of the samples introduced were not labeled as to the length of time they had been in storage. Instead, the panelists were asked to estimate the time for these samples. Prospective panelists who erred by the largest margins were removed from the exercise, and eventually, a panel of the seven most accurate prospective members was established. Three alternates were also retained as substitute panelists in case one or more members of the expert panel might be unavailable.

This training technique was able to suppress the subjective tendencies of the panelists and to bring forth their objective capabilities.

RELIABILITY OF THE EXPERT PANEL

In the study described in the previous section, comprising more than fifty tests, data were obtained to permit an analysis of the accuracy and precision of each panelist as well as of the panel as a whole. It was found that, as individuals, the panelists ranged in accuracy from 71 to 88 percent if they were allowed an error of plus or minus two days. It can be seen that the range of accuracy for the individual panelists is relatively small for sensory data. This information is of importance only for selecting panelists, but is not directly important as a measure of the panel reliability. The important information is that which is obtained for the panel as a whole, because opposite extreme values tend to cancel each other out. Allowing the same error for the panel as a whole, its accuracy was 90 percent. A comparison of the above percentages shows the greater accuracy of panels over that of individuals.

Depending on the expected precision (the number of points in the scale), the accuracy of the panel could be increased to 100 percent by allowing an error of plus or minus three days (this means that the panel would never err by more than three days). When the allowable error was decreased to plus or minus one day, its accuracy fell to 56 percent.

These analyses readily show the fallacy of expecting accurate performance from small sensory panels employing score sheets that have more than just a few points, even when the terms are quantitative. To put it another way, they readily show the need for intensive training of panelists who are expected to differentiate among more than just a few quality levels and do it accurately.

Figure 5.3 shows the response of the expert panel in the experiments described on pages 67–69 compared with the expected response in one

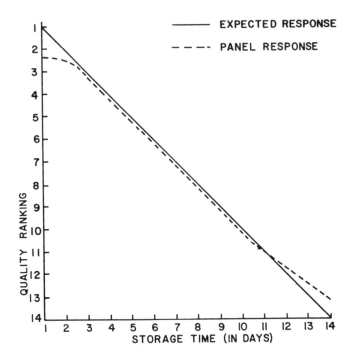

Figure 5.3. An Illustration of Expert Panel Reliability: One Test.

test. The points for the ideal response are, of course, on the line. The points from which the panel response line derives are somewhat scattered about both sides of the line, as is to be expected. The lines are quite close except for the first two days and for the last three days of storage. One interpretation of this discrepancy is that it is purely statistical. There are no values lower than one day of storage and none higher than fourteen. Therefore, there is no possibility of offsetting extreme values, and the diagram shows a panel response line starting near day (or quality ranking) 2 and ending near day (or quality ranking) 13. Still, the striking aspect of this expert panel is its ability to quantify the quality levels of cod fillets over most of the range of storage.

RECOMMENDATIONS FOR FURTHER WORK

The development of the expert panel described above should by no means imply that efforts to improve the reliability of sensory testing are no longer required. We believe that more work can be done in the design of the score sheet, even though the basic principle employed might be continued because of its quantitative value. Another area open to improvement concerns the precision of the panel, because this is the criterion on which the number of points on the score sheet should be based. Finally, an expert panel could be employed to perform more functions. It could be used as the standard for evaluating future objective tests and for reevaluating objective tests already in use. Such a panel could be used to attempt to quantify the organoleptic criteria that are currently used to determine the grades of quality by federal inspectors. These uses would improve the accuracy of the quality testing of seafood of commerce.

REFERENCES

Burgess, J. H. O. 1971. Estimation of fish freshness by dielectric measurement. Report by the Torry Research Station, Aberdeen, Scotland.
Gorga, C., J. D. Kaylor, J. H. Carver, J. M. Mendelsohn, and L. J. Ronsivalli. 1978. The Technological and Economic Feasibility of Assuring Grade A Quality of Seafoods. In-house report of the Gloucester Laboratory, Gloucester, MA.
Gould, E., and J. A. Peters. 1971. *On Testing the Freshness of Frozen Fish.* London, England: Fishing News (Books) Ltd.
Gould, W.A. 1977. *Food Quality Assurance.* Westport, CT: AVI Publishing Co.
Kramer, A., and B. A. Twigg. 1970. *Quality Control for the Food Industry,* Vol. I, 3rd ed. Westport, CT: AVI Publishing Co.
Learson, R. J., and L. J. Ronsivalli. 1969. A new approach for evaluating the quality of fishery products. *Fisheries Industrial Research* 4(7):249–259.

Ronsivalli, L. J., M. S. Schwartz, and J. T. R. Nickerson. 1971. A proposed method for testing the possibility of a botulism hazard in radiation-pasteurized fish. *Isotopes and Radiation Technology* 8(2):211–218.

Wildlife and Fisheries no. 50 Revised 1984. Code of Federal Regulations. Regulations governing processed fishery products and U.S. standards for grades of fishery products. Available from National Seafood Inspection Laboratory, Pascagoula, MS.

CHAPTER

6

Deterioration of Seafood Quality

From a theoretical point of view, freshly caught fish are at their highest quality level. From the moment that fish are slaughtered, their quality begins to deteriorate. Figure 6.1 shows the relationship between the quality of fillets and the time elapsed when the fillets are stored at 0° C (32° F), namely the rate of deterioration of the quality of fresh fillets that are stored at that temperature.

Lean fish have little or no flavor when freshly slaughtered. Even fatty fish at the peak of quality lack a strong flavor. To some individuals accustomed to the presence of varying levels of compounds that are produced in the initial stages of the deterioration of the quality of seafood, freshly slaughtered fish are not as acceptable as fish that have aged for varying times. For example, certain people accustomed to eating fish that have a slight to moderate amount of oxidized fats (rancidity) might not classify freshly slaughtered fish as highest-quality products.

CAUSES OF DETERIORATION OF QUALITY IN FRESH FISH

The factors influencing the rate of spoilage in fresh fish are temperature, level of microbial enzymes produced, and the availability of oxygen. Let us first look at these factors one at a time, and then, since they do not operate in isolation, discuss their combined effect on quality during storage.

72

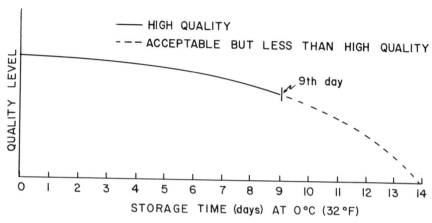

Figure 6.1. Change of Quality in Fish Fillets Stored at 0° C (32° F) Over a Fourteen-Day Period. (From Nickerson and Ronsivalli 1980.)

Temperature

The relationship between temperature and spoilage is well-known (see appendix 4). Studies done by research teams from many countries have consistently found the results that are summarized in table 6.1. It must be pointed out, however, that the data reported in this table relate to fish caught in cold waters. The rate of quality deterioration in fish caught in tropical waters is less well-known, but it appears that the shelf life for those fish, when they are held at comparable temperatures,

Table 6.1 The Shelf Life of Fish Fillets at Selected Temperatures

Temperature			
°C	°F	Shelf Life	High-Quality Shelf Life
26.7	80	1 day	.5 day
15.6	60	2.5 days	1.5 days
5.6	42	6 days	3.5 days
0	32	2 weeks	8–9 days
−1.7	29	3–4 weeks	15.0 days
−12.2	10	c 2 months	c 5.0 weeks
−17.8	0	c 1 year	c 7.0 months
−23.3	−10	c 2 years	c 14.0 months
−28.9	−20	>2 years	> 14.0 months
−40.0	−40	Several years	—

Note: c = about; > = more than.

is longer than that of fish caught in temperate and arctic waters. The data in table 6.1, in brief, indicate that fish can spoil in one day, one week, or in more than one year, depending upon the temperature to which they are exposed.

Enzymes

Enzymes are a class of organic colloids that catalyze chemical reactions; this results in the cleavage or synthesis of compounds. Both the production of enzymes and their effect on the spoilage of fish quality depend on temperature.

There is a considerable body of literature devoted to descriptions of the roles performed by specific microbes (mostly the bacteria) and their enzymes in the spoilage of seafood; since this information is readily available elsewhere, it will not be discussed at length in this book. What is important to recognize is that fresh fish fillets and other fresh seafood lose quality by a variety of chemical reactions involving mainly microbial enzymes that cause the breakdown of proteins. Proteases, the class of enzymes that act on proteins, may behave differently according to the species of bacteria they belong to; that is, they may attack molecules at different sites. Consequently, some enzymes are able to split large protein molecules into polypeptides and peptides, whereas others are able to attack none but the smallest protein components, the amino acids. In general, the objectionable compounds that can be detected in spoiled fish are those produced when amino acids are attacked: ammonia, various sulfides, and other strong offensive end products. One of the compounds reportedly responsible for the typical "fishy" odor is trimethylamine. This odor is an indication that the fish has lost much of its original quality and cannot, therefore, be classified as U.S. Grade A.

At temperatures above freezing, the loss of quality in fish fillets as well as other fresh seafood occurs mainly as a result of the action of enzymes produced by the microbes present on the skin and in the intestinal and respiratory systems of the fish. Since these enzymes are of microbial origin, it has been customary to perform one or more of a variety of microbial tests as an objective means of measuring the quality of the fish. Yet, since a reliable direct relationship between microbial numbers and quality cannot be expected in every instance, the question may be asked, why assess microbes and not enzymes or enzyme activity? This question is not an easy one to address.

Microbial counts and, in particular, the total plate count have long been used as indices of seafood quality. In some cases, standards have been established that include specific microbial limits as criteria for quality, and, therefore, microbial analyses are required. Also, there are demonstrable, though limited, correlations between product quality and microbial counts. Too often, however, analysts forget the indirect and sometimes undefined role of the bacteria or other microbes in the changing qualities of seafood and are therefore at a loss when they are unable to correlate quality with microbial counts.

Since the spoilage of fresh seafood is caused mainly by microbial enzymes (usually bacterial enzymes), the rate of spoilage depends partly on the number of bacteria present. The greater the number, the greater the rate at which enzymes are produced over time. This is the reason counting the number of bacteria is thought to have a bearing on the quality of seafood; but the relationship between bacterial numbers and quality is limited because it appears that, once enzymes have reached a particular level, any increase in the number of bacteria no longer affects the spoilage rate. Most bacteria reproduce by fission: Each bacterium becomes two by dividing itself into two new cells. As shown in figure 6.2, at 0° C (32° F) the generation time for *Pseudomonas fragi,* one of the psychrophilic bacteria reportedly associated with the spoilage

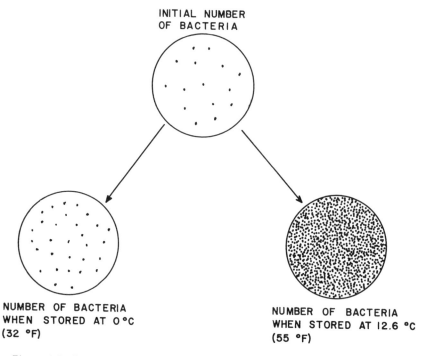

INITIAL NUMBER
OF BACTERIA

NUMBER OF BACTERIA
WHEN STORED AT 0 °C
(32 °F)

NUMBER OF BACTERIA
WHEN STORED AT 12.6 °C
(55 °F)

Figure 6.2. Growth in the Number of P. fragi *on a Microscopic Area of Fish Flesh during the First Twelve Hours of Storage.*

of seafood, is approximately twelve hours. Accordingly, at this temperature the population of this species will double every twelve hours. When this same organism is held at about 13° C (55° F), the generation time is shortened to two hours; hence, at the end of twelve hours the population at the higher temperature will be thirty-two times greater than the population held at the lower temperature. That is, at 0° C (32° F), one bacterium becomes two in twelve hours; at 13° C (55° F), one bacterium becomes two in two hours and sixty-four in twelve hours. Theoretically, at the higher temperature, the rate of enzyme production is also thirty-two times greater than it is at the lower temperature.

Even more precisely, the rate of fish spoilage depends on specific species of bacteria, and there may be several at the beginning stages of spoilage. Most are classified as psychrophiles, which tolerate the cold and can grow readily at low temperatures. Their optimum growth range is 15° C to 25° C (59° F to 77° F). As can be seen in table 6.2, based on different optimum growth temperature ranges, there are two more classifications of bacteria: the mesophiles, which undergo optimum growth in the narrow range of temperatures normal to warm-blooded animals; and the thermophiles, which grow best in the range 50° C to 65.5° C (122° F to 150° F). It should be noted that while the members of these different groupings of bacteria grow best within the ranges given, they also grow at temperatures above and below their respective ranges.

There are two reasons that both the rate of chemical reactions between enzymes and elements of the substrate (fish flesh) and, obviously, the rate at which fish spoil are dependent on the temperature. The first is that within limits, the higher the storage temperature, the faster the growth of bacteria and consequently the larger the amount of enzymes produced within a given time.

The second and most important reason fish spoil more quickly at higher temperatures is that the spoilage reactions, being chemical in nature, follow physical laws of chemistry. The rule of thumb used by chemists is that, within limits, the rate of chemical reactions approxi-

Table 6.2 Classification of Bacteria According to Their Tolerance to Temperature

Classification	Minimum		Optimum		Maximum	
	°C	°F	°C	°F	°C	°F
Psychrophiles	0	32	15–25	59–77	30	86.0
Mesophiles	10	50	37	98.6	43	109.4
Thermophiles	25	77	50–65.5	122–150	85	185.0

mately doubles for every 10° C (18° F) rise in temperature. Thus, a rise in temperature results not only in an increase in the rate of enzyme production, but also in an increase in the rate at which the enzymes react to spoil seafood.

Finally, since microbes have different tolerances to heat, the composition of microbial flora is influenced by temperature, and the manner by which fish spoil is dependent on the composition of the microflora. However, in the range 0° C to 8° C (32° F to 46° F) spoilage does not appear to be significantly affected by changes in the composition of the microflora.

Oxygen

The level of available oxygen has an effect on the composition of microflora. It is under aerobic conditions that the typical spoilage patterns for fresh fish occur; however, lack of oxygen does not imply that the growth of the total microbial population is restricted. Yeasts and facultative bacteria (bacteria that are typically aerobic or typically anaerobic but that have a tolerance for extreme conditions and will continue to grow under some adversities) will grow even when the level of available oxygen is low. When conditions become anaerobic (no available oxygen), only the anaerobes (bacteria requiring the absence of oxygen) will flourish.

Contact of oxygen with fish flesh can be reduced by using gas-impermeable packaging material to protect both fresh and frozen fish. It should be noted, however, that in at least one state, Michigan, it is illegal to enclose a food in a hermetically sealed container, unless the food is to be either frozen or sterilized. The basis for this law centers around a theoretical possibility that botulinum toxin can develop in some unfrozen foods when they are enclosed in gas-impermeable materials because the botulinum bacteria, which are ubiquitous, are strict anaerobes and can grow and metabolize only in the absence of oxygen. There are at least as many scientists who dispute the rationale of this law as those who support it, but until the question is resolved by a comprehensive study of the matter, an argument in favor of the law is that the protection of the public must be preserved in case of doubt. Therefore there is a legal aspect to be considered in selecting packaging materials for seafood.

SOME COMBINED EFFECTS ON QUALITY DURING STORAGE

Temperature, level of enzyme produced, and availability of oxygen operate jointly as spoilage factors, although at times it is difficult to separate the impact that each one may have on the final outcome. In one case involving the use of ionizing radiation to extend the shelf life of fish fillets, half the product was hermetically sealed in oxygen-barrier plastic films, and the other half was packed in oxygen-permeable ones. All of the product was then treated with levels of cobalt-60 radiation in the range of 1 to 3.5 kilograys (100 to 350 kilorads). It was noted that the higher the radiation dose, the greater the bacterial destruction and the longer the shelf life of the product. The cause for the longer shelf life was attributed to the greater destruction of bacteria; however, radiation also destroys enzymes, even though it does so to a far lesser degree. It was also noted that, all else being equal, the products packaged in oxygen-impermeable films had longer shelf lives than products packaged in oxygen-permeable films. The explanation given was that the rate of spoilage in the product packaged in the oxygen-impermeable film was slower because there was an inadequate supply of oxygen to sustain a normal reaction rate.

It was also noted in this study that fish fillets packaged in oxygen-impermeable films developed off-odors and off-flavors that differed from those developed in fillets packaged in oxygen-permeable films. This difference was attributed to variations in the composition of the microflora that developed as a result of both the radiation treatment and the level of availability of oxygen. If the anaerobically packed products were allowed to age for a sufficiently long period, however, it was noted that some of the off-odors and off-flavors were as strong as and similar to those produced when fish fillets are allowed to spoil under aerobic conditions. To explain this result it was theorized that small amounts of preformed bacterial enzymes survived the radiation and engaged in reactions with the minimal amounts of residual oxygen, the scarcity of oxygen and of the reactants accounting for the long time required for the accumulation of detectable amounts of spoilage compounds. At any rate, no evidence was ever produced that allowed us to separate the effects of temperature from those caused by the level of microbial enzymes produced or by the availability of oxygen.

Some Special Cases

In one study there was the unexplained observation of an unexpectedly small difference in the shelf lives of fish fillets handled under extreme sanitary conditions and having a low bacterial contamination; fish fillets handled under normal sanitary conditions and having a normal bacterial contamination; and fish fillets handled under unsanitary conditions and having a high bacterial count—all stored at the same temperature (Huss et al 1974). This result could be explained by the fact that despite the difference in bacterial numbers, and consequently in the amount of enzymes present in each case, the minimum amount of enzymes required to maintain the spoilage rate was attained in each case, and the presence of excess enzymes was irrelevant. The required amount of enzymes was reached first in the fillets having the highest bacterial contamination and last in the fillets having the lowest.

Another study compared the shelf life of fish in which bacterial numbers had been significantly reduced with that of fish in which no attempt was made to reduce the amount of contamination. Disappointingly, there was no significant difference, perhaps for the same reasons as those offered above.

THE DETERIORATION OF QUALITY IN FROZEN FISH

The deterioration of high-quality frozen fish fillets occurs mainly as a result of the action of autolytic enzymes, storage temperature, dehydration, and oxygen.

Autolytic Enzymes

Autolytic enzymes are those inherent in the flesh and are not produced by microbes. For both frozen and fresh fish, then, the agents responsible for the deterioration of quality are enzymes, though from different sources.

Storage Temperature

The rate of deterioration in frozen fish is a direct function of temperature. When the storage temperature is reduced to below freezing, and water in the tissue of the fish flesh starts to freeze, there is a decrease in the water activity. The term *water activity* (A_W) is used to express the degree of availability of water and is defined as the equilibrium relative humidity (RH_e) over 100: namely, $A_W = RH_e/100$. At equilibrium relative humidity, food neither absorbs nor loses water to the air space around it.

Since the development and metabolism of microorganisms occur only in aqueous solutions, the freezing of water in fish tissue at low temperatures immobilizes the organisms and makes it impossible for them to carry out the biological functions that ordinarily result in the spoilage of seafood. Some degradation due to microbial enzymes, already present in the substrate, is theoretically possible, though at much reduced rates. Generally, to sustain their growth, bacteria need higher water activities than do yeast and molds. Many species require water activities of at least 0.96, although there are some species among the halophiles (bacteria that have a high tolerance to salt) that can grow at water activities as low as 0.75. Yeasts grow at lower water activities than those generally required by bacteria, some of them growing at water activities as low as 0.81. Molds can grow at even lower water activities; some are able to metabolize at 0.62.

Although the deterioration of frozen fish fillets is prevented most effectively at the coldest temperatures, there appear to be practical and economic limitations that make it infeasible to store seafood at temperatures below −28.9° C (−20° F). In the storage rooms of a few processors and in some warehouses, frozen fish fillets may be held at −28.9° C (−20° F), but in most cases the temperatures of such rooms are in the range of −23.3° C to −17.8° C (−10° F to 0° F). Unfortunately, at times temperatures are allowed to rise above these levels, and, in other cases, they may be so out of control that they rise to levels that cause some thawing. In the latter cases, the shelf life of the product may not last longer than a few weeks due to the higher rates of enzymatic activity that can be sustained.

Dehydration

One physical phenomenon that contributes to the spoilage of frozen fish fillets is dehydration, which occurs when the amount of water vapor in the air that surrounds the fish is low. When that happens, the seafood loses more water to the air than it gains from it. Both the seafood and the air contain water and therefore each has what is known as vapor pressure. The vapor pressure depends on both the water content of the substance and its temperature. When seafood is exposed to air, water molecules, in the form of vapor (measured as the vapor pressure of the seafood), continually leave the seafood and other water molecules from the air (measured as the vapor pressure of the air) continually condense on the seafood. When the vapor pressures of the seafood and the air are equal, there is neither loss nor gain of moisture by the seafood. When the vapor pressure in the air is lower than in the seafood, there is a continual loss of moisture from the seafood because the number of water molecules leaving the seafood exceeds that of the water molecules from the air condensing on it.

If fillets without a surface protectant are placed in a freezer chamber in order to freeze them, they will lose a considerable amount of water to the surrounding air space. The air space continually loses water to the unit that cools it by the process of condensation, and the water vapor lost by the air is replaced by water from the vapor pressure of the fish.

In earlier times, to prevent dehydration, fish fillets were protected by a heavy ice glaze attained by spraying the frozen fillets with, or dipping them in, water several times, thereby coating them with a succession of several layers of ice. Today, the prevention of dehydration is usually accomplished by prepackaging the product in one of a variety of available freezer wraps. These vary from parchment or other specially prepared papers, through a variety of flexible transparent plastics, to aluminum/plastic laminates. The flexible plastic materials and laminates are superior to ice glazing because the glaze eventually loses its protective ability due to cracking and/or evaporative losses.

In most cases, the package used is made from clear flexible plastics, such as one of the polyolefins. They prevent loss of moisture but allow the passage of gases. Other materials, such as polyester and polyvinylidene chloride, are effective barriers to both water vapor and gases. In either case, these films can prevent dehydration in the product if the package is designed so that the surface of the product is in contact with the package in its entirety. Where areas of the product surface are not in contact with the package, namely, where there are air spaces, ice crystals (frost) will eventually accumulate and water will be irreversibly lost by the product.

When products are poorly protected or when packages with good barrier properties are not tightly sealed, products lose water to such an extent that they may not meet the minimum weight standard. Water lost by the product cannot be reabsorbed: Some of it will adhere to the surface, but the rest will be lost permanently. In surveys conducted by a number of private and government teams to determine the quality of frozen fish fillets available at retail outlets, some fillets were found to be so badly dehydrated that as much as one-third of the weight of the sample was lost (Anon. 1961; Anon. 1965). Dehydration also enhances the rate of rancidification, is associated with discoloration, and may contribute to toughening of the texture.

Oxygen

Experiments in pasteurizing fish fillets with atomic energy and using gas-permeable packaging did not result in the expected extension of shelf life. This failure is believed to have been due to the oxidative reactions of preformed microbial enzymes that could continue to function when oxygen entered the system through the package. Thus, it was confirmed that any significant extension in shelf life required the use of gas-impermeable packaging.

It was assumed that the use of gas-impermeable packaging would apply to nonirradiated frozen fish as well, particularly since it would inhibit the development of rancidity and dehydration. In the last decade, however, scientists at the NMFS Gloucester Laboratory have found evidence that certain species of fish contain a specific enzyme that causes toughening at rates enhanced in the absence of and slowed in the presence of oxygen. It appears that when fish fillets are packaged and held at freezing temperatures we may have one of the following situations:

1. Fillets that contain the enzyme and are packaged in gas-impermeable films will not become rancid, but will toughen.
2. Fillets that contain the enzyme and are packaged in a gas-permeable film will not toughen, but will become rancid.
3. Fillets that do not contain the enzyme and are packaged in a gas-impermeable film will not toughen and will not become rancid.
4. Fillets that do not contain the enzyme and are packaged in a gas-permeable film will not toughen, but will become rancid.

These conflicting conditions cannot be easily resolved when the products contain the enzyme. Yet, it would appear that lean fish that contain this enzyme can be packaged in gas-permeable films even though they might acquire a slight degree of rancidity. Gas-impermeable films are preferable for fatty fish, which do not toughen, but which soon become rancid if exposed to oxygen. A study of the effect of temperatures below $-28.9°$ C ($-20°$ F) on the development of rancidity or toughness in cases 1, 2, and 4, above, is warranted.

THE RATE OF DETERIORATION OF QUALITY

The available data suggest that the rate of loss of quality is slow at first, thereafter accelerating until spoilage is reached. When fish fillets are held at $0°$ C ($32°$ F), as can be seen from the range of time values shown in figure 6.1, the spoilage level is reached in about two weeks; at higher temperatures, the spoilage time range is shorter. For example, when the temperature is raised from $0°$ C ($32°$ F) to about $5.6°$ C ($42°$ F), the time range is halved to about one week. As a key point, the curve of figure 6.1 suggests that even when kept at $0°$ C ($32°$ F), the product remains at a high level of quality for only about eight to nine days. Then, within less than one week, the product quality descends through intermediate levels and ultimately reaches spoilage.

During the early part of storage, the quality changes are difficult to perceive, mainly because the average increment of change is relatively small. During the second week, the average increment of change is somewhat larger and it is correspondingly less difficult to distinguish one level of quality from another. During the first eight to nine days depicted by the curve in figure 6.1, the quality would meet the criteria for U.S. Grade A, the highest of the official grades of seafood quality established by the U.S. government. The latter part of the curve depicts most of the second week of storage at 0° C (32° F) and represents intermediate quality levels. At some point, the limit of acceptable qualities is reached, and, at the end of the curve, the product is defined as spoiled and unacceptable as a food to most consumers.

Since the measurement of quality depends on organoleptic criteria that are measured by sensory as well as objective methods, it is not easy to pinpoint the different levels of quality of the product. In fact, seafood quality is not perceived in exactly the same way by all people. Thus, according to the majority of responses, even when spoilage in a seafood is reached, there is a small but definite probability that some individuals will consider the product to be still acceptable.

PREDICTING SEAFOOD SHELF LIFE

As reported in a number of studies concerning the spoilage of seafood, it has been observed that for any given temperature, time elapse until spoilage can be divided into relatively constant periods: at 0° F (32° C) it is fourteen days, and so on. These findings have prompted at least three research teams from Great Britain (Spencer and Baines 1964), Australia (James and Olley 1971), and the United States (Charm and Ronsivalli 1967) to investigate the possibility of predicting the effect of temperature on the spoilage rate of fish fillets. All three teams have come to the same conclusion: Shelf life can be predicted for any given storage temperature in the range 0° C to 8° C (32° F to 46° F), using the Arrhenius equation or graphical methods developed by these researchers. Spencer and Baines found that the range of temperatures over which the shelf life could be predicted was higher than the ranges of temperatures found by the other two teams. The Australian team developed a temperature-function integrator and multipoint telemeter to predict the shelf life of seafoods.

A United States team (Charm et al 1972), designed a graph from which a slide rule type of predictor was developed (Ronsivalli et al 1973). It can integrate the increments of shelf life that have been consumed (shelf life consumption being a function of time and temperature), and yields an estimate of the shelf life that remains for any temperature in the range mentioned above. (It should be noted that the temperature scale shown in the device is in Fahrenheit only.) This device has the advantage of low cost, ease of portability, and ease of operation, and it can be useful to anyone engaged in the handling of fresh fish.

In one outstanding example of its utility, the shelf life predictor was used to obtain the answer to a question posed by a potential exporter of seafood to Milan, Italy. The exporter needed to know whether his product would remain of high quality for two days after it reached Milan, provided that the product was held at 0° C (32° F). In other words, he wanted to know how much shelf life would be used during production and transportation, in order to determine whether there would be at least two shelf life days left once the product reached its destination. The predictor indicated that the shipment could be made with confidence. Without the shelf life predictor, this particular problem could not be solved except by trial and error, a costly and time-consuming process.

With the current availability of computers, the device can be transformed into a computer program in order to simplify its mode of operation and extend the range of its applications. It should also be noted that since the shelf life predictor was derived from experiments with cod fillets, it may have to be calibrated for application to other species.

REFERENCES

Anon. 1961. Frozen fried fish sticks. *Consumer Reports* 26(2):80–83.

Anon. 1965. Frozen fried fish portions. *Consumer Reports* 30(5):235–237.

Charm, S. E., and L. J. Ronsivalli. 1967. Effect of processing on microbial flora. *Food Technol.* 21(5):60–64.

Charm, S. E., R. J. Learson, L. J. Ronsivalli, and M. Schwartz. 1972. Organoleptic technique predicts refrigeration shelf life of fish. *Food Technol.* 26(7):65–68.

Fields, M. L. 1979. *Fundamentals of Food Microbiology*. Westport, CT: AVI Publishing Co.

Gorga, C., and L. J. Ronsivalli. 1983. Quality control and quality assurance: Getting the difference straight. *Infofish Marketing Digest*. 4:32–34.

Huss, H. H., D. Dalsgaard, R. Hansen, H. Ladefoged, A. Pedersen, and L. Zittan. 1974. The influence of hygiene in catch handling on the storage life of iced cod and plaice. *Journal of Food Technol.* 9:213–221.

James, D. G., and J. Olley. 1971. Spoilage of shark. *Australian Fisheries* 30(4):11.

Licciardello, J. J., L. J. Ronsivalli, and J. W. Slavin. 1967. Effect of oxygen tension on the spoilage microflora of irradiated and nonirradiated haddock fillets. *J. Appl. Bact.* 30(1):239–245.

Liston, J. 1982. Recent advances in the chemistry of iced fish spoilage. In *Chemistry and Biochemistry of Marine Food Products*, R. E. Martin, G. J. Flick, C. H. Hebard, and D. R. Ward (Editors). Westport, CT: AVI Publishing Co.

Nickerson, J. T. R., and A. J. Sinskey. 1974. *Microbiology of Foods and Food Processing*. New York, NY: Elsevier Publishing Co.

Racicot, L. D., R. C. Lunstrom, K. A. Wilhelm, E. M. Ravesi, and J. J. Licciardello. 1984. Effect of oxidizing and reducing agents on trimethylamine n-oxide-demethylase activity in red hake muscle. *J. Agri. and Food Chem.* 32(3):459–464.

Ronsivalli, L. J., R. J. Learson, and S. E. Charm. 1973. Slide rule for predicting shelf life of cod. *Mar. Fish. Rev.* 35(7):34–36.

Slavin, J. W., J. A. Peters, and S. R. Pottinger. 1958. Studies on a jacketed cold-storage room. *Food Technol.* 12(11):602–609.

Spencer, R., and C. R. Baines. 1964. The effect of temperature on the spoilage of wet white fish. *Food Technol.* 18:769.

7

Quality Assurance—Not Simply Quality Control

During the first half of the twentieth century, the label "made in Japan" was tantamount to a warning not to buy—such was the stigma earned by Japan for its inattentiveness to the importance of the quality of its manufactured products. The only selling feature of Japanese goods was their lower price, but this was evidently not enough to establish Japan as a significant factor in international trade. During the second half of the twentieth century, however, there gradually occurred a dramatic transformation in the economic status of Japan and in its impact on international trade. To many, this transformation was either unexplainable, or rationalized and supported by a variety of theories. Actually, what happened is becoming exceedingly clear.

Once the Japanese became convinced that their economic plight was due to the old image of products made in Japan, their course of corrective action became self-evident. Their proposed philosophy was to produce items of such high quality as to establish a leading international reputation. Both their analysis and the course that they set proved to be correct, and within two decades or so Japan had ascended to a status of gigantic proportions in international trade, riding the crest of an enviable wave of exports of the highest-quality automobiles, automobile parts, cameras, appliances, electronic equipment, and many other products.

Is there any stronger evidence of the importance of quality in today's markets? Doubts on this score are so few that the issue has been generalized to a very high and broad level. It has, in fact, been observed that we are in "The Third Wave" of human activity (Toffler, 1980). The first wave was represented by the agricultural revolution, a revolution that gradually freed many people from the slavery of working for physical sustenance alone. With the introduction of machinery and large-scale mass production we entered the second wave, namely the industrial revolution. The third and current wave is represented by too many innovations and characteristics to be mentioned here, but one certainly is the application of measures to assure the quality of products and services. In this wave, unlike the previous ones, the United States is lagging behind other countries and especially behind Japan. This situation is more distressing because much credit for Japan's phenomenal economic growth is attributed to the quality control measures designed by an American, Dr. W. Edwards Deming.

In brief, although the reputation of some U.S. products has suffered a setback over the past couple of decades, a reversal is not impossible. Few industries have so many opportunities for improvement as the seafood industry. The pivot in this endeavor is the need to assure the quality of seafood at the point of consumption, where quality ultimately matters. But first, let us define the problem.

QUALITY CONTROL VERSUS QUALITY ASSURANCE

Quality control is an internal matter to each element of the industry, and many manuals have been devoted to the subject. Quality assurance, on the other hand, is so rarely discussed outside of completely integrated industries that some authors have declared the two to be interchangeable terms. But quality control and quality assurance are not equivalent. Indeed, there is a substantial difference between them. *Webster's New Collegiate Dictionary* defines *quality control* as "an aggregate of activities (as design analysis and statistical sampling with inspection for defects) designed to insure adequate quality in manufactured products." It is both relevant and interesting to note that the dictionary does not define the term *quality assurance*. Yet, while *control* is defined as regulation, domination, restraint, manipulation, and so on, *assurance* is defined as guarantee, insurance, sureness, warranty. Thus the dictionary definitely implies that there is a difference between the two terms.

The Technical Difference

Quality control is a set of measures taken within each firm; quality assurance is a set of measures taken across an entire industry. While

quality control is the responsibility of each element in the chain of production and distribution of any product, quality assurance is the responsibility of the industry as a whole. Quality control measures can and
do vary from element to element of the industry. Quality assurance
is an industry-wide *common* pledge: namely, to assure quality of the
product at the point of consumption. In technical jargon, the difference
is between a "parts" and "systems" approach. An authoritative report
concerning the application of systems theory to the food industry was
issued in 1985 by the National Research Council of the United States. It
is entitled *An Evaluation of the Role of Microbiological Criteria for Foods and
Food Ingredients*. The report endorses a program of microbiological control called the Hazard Analysis Critical Control Point system (HAACP).

In brief, the key difference between the parts and the systems
approaches is one of time span: Quality control involves activities that
tend to prevent or minimize deterioration of quality while the product
is within the premises of the firm; quality assurance, on the other hand,
involves activities that maintain the expected quality level—U.S. Grade
A quality, in the case of seafood—up to the moment the product is
finally used by the consumer. The key technical difference is that quality
assurance includes quality control measures, but requires much more.

The practical consequences of this difference between the two
approaches are so significant that they hold the key to why, after so
many years and great efforts at various levels of endeavor, prior to the
cooperative government/industry effort in the area of quality assurance
that was run by the Gloucester Laboratory, not the slightest permanent
improvement had been recorded in the quality of seafood available to
consumers in supermarket retail outlets.

QUALITY ASSURANCE AT THE SUPERMARKET LEVEL

It is the poor quality of seafood sold at supermarkets that has given
the seafood industry as a whole a bad name. And indeed it is because
of the relatively large volumes sold at supermarkets that there is much
concern about the need to improve seafood quality. In contrast, however, seafood bought in retail specialty stores and consumed in restaurants is generally of high quality.

There are a number of historical, technical, and economic reasons
most operators of supermarkets are not as concerned with the quality
assurance of seafood as, for instance, are restaurateurs. In most cases, supermarket operators are not knowledgeable about the quality of
seafood. Besides, seafood is such a relatively small fraction of sales
of protein foods in general that there seems to be no apparent economic motive to undertake the efforts necessary to assure its quality.

In addition, the generally poor quality of seafood in supermarkets does not affect the reputation of the supermarket operator but that of the processor whose name is often identified on the package. Most important, the fact that the problem of poor-quality seafood is discovered at the supermarket level does not necessarily imply that the problem was created there. For the moment, our concern centers around supermarkets, not only because this is where the problem of poor-quality seafood is most concentrated and most visible, but especially because this condition is an outstanding example of the ineffectiveness of quality control measures taken without consideration for quality assurance.

THE INEFFECTIVENESS OF QUALITY CONTROL ALONE

In recent decades, various organizations have taken numerous surveys of the quality of seafood available in supermarkets and they have all reported dismaying results. In one such survey, published more than twenty years ago, *Consumer Reports* was prompted to write:

> One likely reason for this country's low consumption of seafoods—which was held at 10 or 11 pounds a year per person for more than a generation—is that most people seldom get to taste the sweet, delicate flavor of fresh-caught fish. It's probable that this plentiful food, rich in protein, vitamins, and minerals and relatively inexpensive, goes a-begging because, by the time it reaches the dinner table, it has usually attained an age and condition warranting its religious connotation as a penance food (Anon. 1965, 235).

We quote from this relatively old report because there are several truths in it that are still valid. One is that fresh-caught fish are indeed a delicacy. It is safe to say that hundreds of different species of fish have delicate, desirable flavors if they are eaten soon after they are caught. Another outstanding set of truths is that the best gustatory experiences with seafood take place in restaurants; that specialty seafood markets sell products of relatively good quality; and that most supermarkets in the United States continue to sell both fresh and frozen seafood products that, for the most part, are of poor quality.

What the above quotation or, for that matter, most surveys and technical analyses do not reveal is *where* the problem of poor-quality seafood originates. Let us assume that all other operators upstream have strictly adhered to their quality control measures. Have those efforts been sufficient to provide the consumer with a high-quality product? Evidently not. Quality control measures alone are ineffective.

Nor is it to be assumed that efforts directed at quality control have been slight. Innumerable studies have been aimed at parts of the system, such as innovative harvesting techniques, temperature control techniques, microbial control measures, and nearly every other conceivable aspect of quality control. These studies invariably recommend a set of quality control measures for only a particular element in the handling chain. And there is the crux of the matter. Generally, efforts to improve the quality of seafood have been directed at isolated elements of the chain, which starts at the fishing vessel and ends at the retail market. But the system as a whole has been neglected. Thus, the goal of quality assurance is not quite reached. It seems that quality controls, at least as practiced in that segment of the seafood industry that sells its production through supermarkets, have not resulted in quality assurance—the guarantee of a quality product at the moment of consumption.

QUALITY ASSURANCE

"By the time that 'Columbus sailed the ocean blue, in fourteen hundred and ninety-two'," reported Gerald Robert Watkin, Chief Inspector of the Worshipful Company of Fishmongers of the City of London, "the Fishmongers Company had been in existence for over two hundred years" (Watkin 1980, 58). Its work involved keeping out the "foreigners" from south of the Thames and, with royal approval, fixing prices. He concluded: "When prices are fixed, so must be the quality and, therefore, each of the victualling [food and drink] guilds had power to inspect and condemn" (ibid., 58).

"Mr Ogvind Lie, of Frionor Norsk Frossenfish [_sic_], Oslo . . . quoted the following Norwegian quality programme _anno 1444_: 'Stockfish quality grading shall hereafter be mandatory. . . . Each one who is doing otherwise . . . has lost the ownership of the stockfish in question, and this shall thereafter belong to the kingdom. . . . And the grading men should have for their strive, one half fish for every hundred and twenty they have been grading'" (Phillips 1980, FS3).

It is evident, then, that while the use of the term _quality assurance_ appears to be relatively new in industrial parlance, the goal of assuring the quality of seafood to the consumer is ancient—and reachable. The discussion of more recent developments concerning the achievement of this goal is divided into two parts. The first is on countries that have earned a deserved reputation for their high-quality seafood. These countries have, in effect, set standards against which, in a world of intense international trade competition, other countries have to be measured. The second part is on countries that have lately shown an increasing awareness of the need to improve the quality of their seafood.

Quality, A Continuing Effort.

In the process, we should not only gain confidence in the feasibility of a quality assurance program but even pick up some valuable insights into the actual means for its implementation. The key point is that lack of assured quality to consumers is the single most important impediment to the growth of seafood consumption and the growth of the seafood industry as a whole.

Countries with Reputations for High-Quality Seafood

Some countries—Norway, Denmark, Poland, Iceland, and Japan, among others—are recognized worldwide for the consistency of the high quality of their seafood. In Norway, for example, fish destined to be filleted are promptly bled, gutted, washed, and carefully stored and unloaded. Vessels are limited to short hauls, and processing plants produce top quality fillets. Regulations are extremely detailed, down to such minutiae as the incandescence and placement of lights over filleting tables. Fish are frozen preferably before rigor mortis sets in or while still in rigor. Freezing plants make sure that all large fish, such as tuna, porbeagle, or skate, are chilled in ice to at least 4° C (39.2° F) before the freezing begins. Packing and freezing is done in such a way that the products are cooled to a temperature of −15° C (5° F) or colder within twenty-four hours after they have been placed in a freezer.

Seafood products from Iceland are renowned for their high quality. Among the factors that assure this result are the proximity of the fishing grounds to the processing plants and the attitude about quality that exists throughout the industry. There is also a reverse side to this coin, however. While strict control measures are usually required to achieve high-quality products, care must be exercised that over-regulation does not interfere with productivity and efficiency. Iceland has run into this problem, but appears to be on its way to correcting excesses of regulation.

In Japan, fish destined for the fresh seafood trade are not more than three days out of the water when they are landed in port. Although the average trip length for the off-shore fleet is forty days, a system of carrier boats picks up the fresh catch every three days. Off-shore trawlers are also equipped with a plate-freezer, ice-maker, and desalinator. Equipment and gear aboard vessels are up-to-date; crew accommodations are excellent and clean. Fish destined for the fresh market are packed on board in ice in Styrofoam boxes that are used only once.

Recognizing the Need to Improve Quality of Seafood

Of the larger seafood-producing nations, the United States and Canada have recognized the need to improve the image that they currently have in the international arena. Both countries, as well as others, have demonstrated the benefits of assured quality and have made strong commitments to continue on this course.

One of the major contributions by the United States to assure the quality of seafood was a cooperative effort involving the NMFS and a few members of the seafood industry. By early 1980 it was found that the awareness of the need for quality assurance had become widespread in the United States. The Maine Department of Marine Resources and the Maine Development Foundation combined to develop a quality control program for adoption by the industry. The Rhode Island Seafood Council announced its intention to implement a quality program that would initially be paying participating vessels a premium price for fish to be processed and distributed under the U.S. Grade A label. The New

England Fisheries Development Foundation has also carried out similar experiments in New Bedford, Massachusetts.

Similar cases are reported in other areas of the country. A processor from Seattle, Washington, for instance, buys everything fishermen catch and pays them about double the price on most species, but insists on the following standards:

> All fish must be caught on hook and line. No fish which comes over the rail dead will be sold as fresh fish. All fish will be bled immediately while still alive. All fish, with the exception of rockfish, which are sold whole and ungutted, will be cleaned within two hours after bleeding. All fish must be iced immediately after cleaning and sorted by size and species into 100 liter totes. The totes are taken off the vessel no later than the next day and trucked to Seattle directly to buyers (Fitzgerald, 1980, 21 [italics in original]).

In brief, it is by insisting on these practices that he can guarantee the quality of his product. In particular, he guarantees that his fish is not older than three days. He charges up to twice the going price and his buyers cannot get enough of his fish.

If these cases appear to be too particular and individual, there are others that unmistakably prove the validity of our claim. On a nationwide scale, restaurants and fast-food chains, the McDonald's chain in particular, are readily demonstrating that it is quite feasible to assure high-quality seafood, even frozen seafood.

In the United States, quality has come to be recognized as the key to opening up new markets for fishery products. Indeed, the need for an official assessment of the quality of seafood products has ultimately been recognized at the highest levels of the U.S. government. At the beginning of 1981, it was widely reported that the General Accounting Office (GAO) had asked NMFS to conduct a survey to document the extent to which quality defects exist in seafood products produced in the United States.

The focus of attention in Canada has become the preparation of official standards published by the Department of Fisheries and Oceans in a booklet entitled *Quality Excellence in the 80's*. The overall policy is that "all sectors of the industry must cooperate to ensure that the Canadian name on a product brings automatic recognition of top quality on the markets of the world." (ibid, 1) The implementation of this policy involves a detailed and scheduled program of vessel certification; quality protection on board; dockside grading; unloading, dockside handling, and transportation to plants; improved quality control in processing plants; final product grade standards; and advice on handling and processing practices.

Official efforts to improve quality have been paralleled by private industry efforts. In 1980 National Sea Products Limited and B.H. Nickerson and Sons Limited announced the formation of a jointly owned research and development company, Fisheries Resource Development Limited. The central area of involvement for this enterprise was planned to be quality improvement and product development. The importance of high quality has also been firmly supported by the Fisheries Council of Canada, whose chairman is on record stating the following: "It is my strong conviction that quality enhancement is an essential key to market growth at home and abroad" (McLean 1980, 5). The president of the Fisheries Association of Newfoundland and Labrador is also on record stating that quality is a key to the future of the industry.

In England the administrative structure of the official quality control program is under two authorities. Outside the City of London, this responsibility is carried out by the Environmental Health Inspector; in London, it is still carried out by the Worshipful Company of Fishmongers, mentioned earlier. The company is a private organization administered by a chief inspector and two assistants (fishmeters). In addition to quality control, the company also performs relevant studies; is a funding source for other fisheries research organizations; and carries on an extensive training program.

It is the official policy in the USSR to have fish of the highest possible quality. The minister of fisheries has stated that the number one task for the fishermen today is "the quality of the product" (Ishkov 1975, 52).

In Australia the initiative to assure seafood quality was taken by a supermarket chain, G. J. Coles and Co., Ltd. After tentatively test marketing branded, prepacked fish, chilled at 0° C to 2° C (32° F to 35.6° F) and supplied by an independent processor, in a handful of stores, a network of production centers and distribution systems covering most of Australia was established in little over two years, and six new processing and packing plants were built. Even in inland areas remote from sources of fish, the demand was found to be higher than average, and the added cost to assure quality was readily paid by consumers.

The adoption of the 200-mile zone legislation in New Zealand—just as in other areas of the world—brought with it substantial investment in new vessels and processing plants, and consequently difficulties in marketing the increased catch. A study commissioned by the Department of Trade and Industry found that quality assurance offers the best opportunity for market expansion.

At the third International Seafood Conference (1980) organized in Rome by the Food and Agriculture Organization (FAO) of the United Nations, three major points relevant to this discussion were raised. First, it was noted that world exports of fisheries products had increased considerably in the decade of the 1970s (from $3,392 M to $11,170 M), with the share of developing countries increasing at a rate one-third higher than that of developed countries (from $1,000 to $3,800 M). Second, it was stressed that there are problems, faced especially by developing countries, related to product quality, prices, and insufficient and/or costly shipping facilities. Third, it was pointed out that fish must be made "a chic experience" and should not be sold just as food (this point has been confirmed and stressed by an illuminating in-depth survey of consumer reactions in the United States, the Miklos Report).

The points raised at this FAO conference might serve as the backbone for a few comments to this brief review of a sample of past and present quality assurance programs—and attempts at quality assurance—all over the world. First, the recent worldwide growth in demand for seafood represents an opportunity to be exploited by the United States as well as by many other countries. Second, industry worldwide has many obstacles to overcome. Third, making fish "a chic experience"—namely, not only producing high-quality seafood, but also making it look like a high-quality product—appears to be the method through which obstacles can be overcome and the potential benefit of the seafood industry to humankind can be achieved.

We believe that the essential prerequisite for the achievement of these goals is the development of a quality assurance program, and will now proceed to elaborate on the elements of the program.

ESSENTIAL ELEMENTS OF A QUALITY ASSURANCE PROGRAM

The whole of a quality assurance program can be reduced to one recommendation. Each and every participant in the handling chain should strictly adhere to these two simple rules:

1. Make sure that the seafood entering the system has high enough reserve quality to last until the moment of consumption.
2. Handle the product in such a way as to minimize the loss of quality.

While the following chapters will provide detailed explanations of why and how these rules are to be followed, in the remainder of this chapter, we wish to highlight a number of factors we believe are essential to the success of the quality assurance program:

Pledge of quality assurance
Sanitation
Product safety
Quality control
Strict adherence to a timetable
Separation of the catch at landing
Inspection
Grade labeling
Planning
Program coordination
High initial reserve quality

Pledge of Quality Assurance

The pledge of quality assurance takes a variety of forms. The essential components of this pledge, however, are simple: first, have the strictest possible concern for all aspects of sanitation and product safety; second, maintain the strictest possible adherence to quality control measures; third, establish a coordinated program of activities with all other elements in the chain of production and distribution of the product; and finally, replace or reimburse the buyer for any product that is below the highest possible quality level.

Sanitation and Product Safety

Sanitation and product safety are separate but interrelated factors. Provisions must be made to protect the consumer against food-poisoning incidents or any situations that evoke even mild concern about the safety and wholesomeness of seafood.

The only procedure that has legal standing to provide the assurance of safety and wholesomeness of the product is the use of the USDC Inspection Service. This is one of the major reasons this service was used by all quality assurance projects run by the Gloucester Laboratory.

Quality Control

What has been said about sanitation and product safety could be repeated almost verbatim for quality control measures. We cannot overemphasize the importance of these measures. They will be discussed in the next three chapters. Here we simply wish to emphasize the point that presumably has already been made abundantly clear: Quality control is a necessary component of quality assurance, but quality control measures alone do not necessarily lead to quality assurance.

Strict Adherence to a Timetable

Since the assurance of the quality of seafood depends on the performance of a number of participants in the chain of distribution, the degree of quality that can be assured is only the level that can be achieved by the least effective link in the chain. Therefore, regardless of the efforts that might be expended by any or most of the seafood-handling participants, quality will not be assured unless *all* of the elements are effectively coordinated to fulfill prescribed roles.

Our attention must therefore be focused on the major variables on which the assurance of quality depend: The first is temperature and the second, time (see chapter 6 for a more complete discussion of these issues). The linkage between the two determines seafood shelf life, which is usually defined as the maximum period of storage during which the food remains acceptable to consumers (see chapters 5 and 6). We can immediately begin to see that while we have little control of one of these variables, namely time, we can manipulate the other, temperature.

Manipulation of Temperature. In the United States fresh fish is not held at a constant temperature throughout distribution. Instead it may be held at one temperature aboard the vessel, at another in the processing plant, and at still different temperatures during transit, bulk storage, at the retail outlet, and in the home. So, as seafood is put through the distribution system, it constantly loses quality, but at variable rates, depending on the temperatures to which it is exposed.

Table 7.1 provides an indication of the relationship between the holding temperature and the expected shelf life of the product (Gorga and Ronsivalli 1983). These data are derived from experiments involving gadoid species and may not necessarily apply to other species. In addition to species differences, other variables such as season and geographic latitude may exert an influence on what occurs. Still, these data are as reliable as any set of values that can be assembled for the intended purpose, which is to show the effect of the storage temperature on the shelf life of seafood.

In particular, it can be seen from this table that if fish are held at ambient temperatures during the summer months, it will probably become inedible within a period of only one to two days. If freshly slaughtered fish is frozen and stored at about $-28.9°$ C ($-20°$ F), it will remain at a high level of quality for more than one year; it will remain at lower

Table 7.1   The Shelf Life of Fish Fillets at Selected Temperatures

Temperature			
°C	°F	Shelf Life	High-Quality Shelf Life
26.7	80	1.0 day	.5 day
15.6	60	2.5 days	1.5 days
5.6	42	6.0 days	3.5 days
0	32	2.0 weeks	8–9 days
−1.7	29	3–4 weeks	15.0 days
−12.2	10	c 2.0 months	c 5.0 weeks
−17.8	0	c 1.0 year	c 7.0 months
−23.3	−10	c 2.0 years	c 14.0 months
−28.9	−20	>2.0 years	> 14.0 months
−40.0	−40	Several years	—

Note: c = about; > = more than.

levels of quality for a much longer time. Thus, all the eating quality of seafood is used up in one to two days or one year, depending on the temperature at which it is held.

Quite apart from these extreme values, of special interest are the middle ranges. Those that apply to fresh fish tell us that products that are stowed in ice, or at the temperature of melting ice, remain of U.S. Grade A quality for a period of eight to nine days. While their organoleptic quality deteriorates after that, they are still acceptable until about the end of the second week of storage. From then on, most consumers will probably consider them to be of unacceptable quality. In other words, in accordance with the best available data, fresh fish fillets remain at U.S. Grade A quality for eight to nine days.

It is this bit of information that makes us issue again the recommendation so frequently stated throughout this book: Keep fresh fish constantly at 0° C (32° F). More important in the present context, this bit of information helps us suggest how to manipulate even what seems to be beyond our control, namely time. True, we cannot actually control the passage of time, but we can control what we do while time goes by from capture of the fish to its consumption. Specifically, we can develop a schedule of activities to be strictly adhered to by all participants. A proposed schedule is given in table 7.2 and is discussed below.

Allocation of Time: Fresh Fish Schedule. Since fresh fish fillets held at 0° C (32° F) cannot be older than nine days when consumed if they are to retain their U.S. Grade A quality level, limits have to be placed on the times that the product can be held at each stage of the distribution chain. One effective procedure that can be used to arrive at these time

Table 7.2   Time Allowed Each Element in the Distribution of Fish Fillets

	Maximum Time Product Can be Held	
Distribution Elements	At 0°C (32°F)	At −17.8°C (0°F)
Vessel	2 days	7 days
Processing plant	1 day	1–2 days
Warehouse		About 6 months
Retail outlet	5 days	About 3 months
Home	1 day	About 3 months
Total time	9 days	About 1 year

allocations is to start at the moment of consumption and work backwards to the moment when the fish are harvested or slaughtered.

If we consider fresh fish sold at supermarkets and we allow that the consumer may hold the product in the refrigerator (preferably, also in ice) for up to one day after the fish is purchased, then the product must be sold by the retailer by the eighth day of the product's shelf life. If we allow the retailer five days to sell the product, then the processor must deliver the product to the retailer by the third day. If it takes the processor one day to receive, process, and ship his product, then he cannot accept fish more than two days old. If the processor obtains his seafood supply directly from the fishermen, then the fishermen must bring their catch to port within two days from when it was caught.

If the fishing trips are two days or less in duration, then all of the catch is eligible for distribution as fresh fish in the quality assurance program that includes supermarkets. If the trips are longer than two days, then only the top of the catch (fish two days old or less at the time of landing) is eligible for distribution as fresh fish in supermarkets. Fish to be distributed through restaurants, specialty stores, and other outlets that rapidly sell it to the consumer may be older than two days when landed, but in no case should fresh fish older than eight days be allowed to reach the consumer, provided that the fish have been held at 0° C (32° F) during the eight days. It is strict adherence to such a timetable that, in conjunction with other measures, can eventually assure success to the program.

Possible Adjustments. Whenever an extra day is allowed to one of the elements in the fresh fish handling chain, that day must be subtracted from one of the other elements, because in no case can the total of holding days by all elements be increased, unless it can be demonstrably justified. It is possible that the total holding time for fresh U.S. Grade A quality fish may be found to be inappropriate for certain situations. In that case, an adjustment will have to be made in the time allocations.

Obviously, within any given set of constraints, each one of the elements in the fresh fish distribution chain would want to increase its time allocation. In particular, it might appear that the retailer has an excessive allocation of time relative to the other elements. That allocation, however, is not as large as it seems. In reality, the retailer has only four days in which to sell his fresh product, because on the fifth day he must either risk taking a total loss on the unsold product, or mark the product down for sale at a reduced profit or no profit (even this action can result in financial loss); alternatively, he can expend effort and energy to rewrap the product, freeze it, and sell it as frozen product at a lower price.

The retailer can either buy less or he can buy as much as he estimates he can sell within the allotted time. If he buys less, he reduces the degree of financial risk to which he exposes himself, but he does place an artificial obstacle to potential growth in demand. A lack of sufficient and consistent supply in the face of a growing demand tends to suppress the growth of the demand and thus reduces potential production levels and profits for the entire industry. In fact, in order to obtain a real measure of both the demand and any upward trend in the demand the retailer must initially buy more than he estimates he can sell in an allotted time so he will never run out of supply; and, if he should remain with a surplus, he can freeze the product while it is still at U.S. Grade A level of quality. Buying less is in conflict with both the economic aims of the quality assurance program and the consumer's demand for high-quality seafood. This is the reasoning that brought us to the conclusion that the high allocation of time for the retailer only appears to be excessive.

Allocation of Time: Frozen Fish Schedule. Using the same technique that permits us to arrive at a schedule for the fresh fish distribution chain, we can develop maximum times for each member of the frozen fish distribution chain. Given, in accordance with table 7.1, that frozen seafood has a shelf life of about one year at $-17.8°$ C ($0°$ F) and assuming that a consumer might hold the product for up to three months in the home freezer, the retailer should not sell frozen seafood that is more than nine months old. Consequently, if the retailer is allowed a period of up to three months to sell the product, the processor or broker who supplies the retailer should not ship out frozen seafood that is more than six months old. However, if the product has been stored at temperatures colder than $-17.8°$ C ($0°$ F), then longer times are possible.

Within reasonable limits, the length of time allotted to the processor is variable and depends on the knowledge of the history of the specific product. The processor who freezes the fish must freeze it before the fish is nine days out of the water (that is, on the eighth day at most). If the processor requires one day to process the fish (including freezing it), he can process fresh fish that is up to seven days out of the water. If he requires two days for the process, then the fish cannot be more than six days out of the water, and so on. Fishermen have up to seven days to bring in any part of their catch that is going to be frozen and distributed as a frozen product.

A processor who reprocesses frozen products (e.g., makes frozen fish sticks from frozen fish blocks) without thawing the product during any part of the operations has as much time as he reasonably needs to perform his functions.

The Technology Variable. Among all other variables that are implicit in the above discussion is the assumption that currently predominant forms of technology are being used. In fact, forms of technology other than the ones implicitly stipulated above are in use today and still others may come into use in the future. For instance, one of the ways to extend the holding time for fresh seafood is to lower the holding temperature: Lowering the temperature of fresh fish to the superchill range ($-3°$ C to $-1°$ C or $26.6°$ F to $30.2°$ F) will extend the shelf life of the product substantially, with the degree of extension depending on the temperature. Another demonstrated way to increase the fresh fish shelf life is to treat it with a pasteurizing dose of radiation, which, at suggested levels, can approximately double the shelf life; but approval for the use of this technology to preserve seafood has not been given by the Food and Drug Administration (FDA). The point we wish to emphasize is that the suggested timetables for fresh and frozen fish are valid within the limits of prevailing technologies.

Separation of the Catch at Landing

If one misinterprets the implications of the timetables suggested above, one might conclude that they are too stringent and that therefore they might never be used. It might, for instance, be said that there is not enough fresh fish landed to satisfy the requirements of the quality assurance program. Calculations done for New England landings on data furnished by the Analytical Services Branch, Northeast Region of NMFS, tend to dispel these fears. Summary figures are reported in table 7.3, and these indicate that about 50 percent of the catch is qualified to enter the line of distribution that reaches the supermarkets as fresh fish.

Table 7.3   Estimated New England
Landings of Fish, Two Days Old or
Less, by GRT Class, 1980

GRT Class*	Estimated 2-Day-Old or Less Landings (mill. lbs.)
5–50	61.90
51–125	65.20
125 +	25.66
	152.76**

*As defined by the U.S. government, GRT
stands for Gross Registered Tonnage, namely the
internal cubic capacity of fishing vessels. This
measure is expressed in tons of 100 cubic feet.
**This amount represent 48% of total landings.
The remaining 52% represents that proportion of
landings that had been out of the water for more
than two days.

The question, of course, then becomes, What does one do with the
rest of the catch? One can sell the rest of the catch as fresh fish—
and high-quality fresh fish, at that—provided one sells it to a nearby
restaurant or a specialty store that has a fast turnover. Otherwise, this
fish must be frozen, salted, or preserved in some other fashion.

The key to these decisions lies in a detailed knowledge of each batch
of the catch, especially knowledge relating to time and temperatures.
The catch must accordingly be separated at landing, and each batch
introduced into the most appropriate channel of distribution. To do
otherwise is to court disaster: The risk is that quality will not be assured,
the program will fail, and the consumer will be utterly dissatisfied.

Inspection

It is not widely known, especially by consumers, that of all meat
foods produced in the United States, only seafood is not under the
rigid mandatory federal inspection system. Thus, there are no federal
inspectors aboard fishing vessels to oversee the harvesting, preprocess-
ing (such as eviscerating, shucking, etc.), and stowage of fish and
shellfish, whereas federal inspectors do oversee the slaughtering, processing, and handling of red meats and poultry. There is no mandatory
federal inspection in seafood processing plants (except for the voluntary federal inspection services of the National Marine Fisheries Ser-

USDC Inspector at Work.

vice), whereas all plants that process red meats and poultry are under mandatory federal inspection. Therefore, even when a quality control program can be defined for fishing vessels or for seafood processing plants, what assurance can there be that it will be carried out when there are no federal inspectors to see it through? Many seafood experts as well as many critics of the seafood industry have concluded that the lack of mandatory federal inspection is the root cause of the generally poor quality of seafood available at supermarkets.

Still other seafood experts contend that even mandatory inspection of seafood aboard vessels, in processing plants, and in warehouses may not be effective unless the inspection is extended to the retail outlets. The reason is that some ignore display time limitations; many do not know when a seafood product should be discarded (not marked down, but discarded); and most operators of food markets do not always maintain adequate storage temperatures. The quality of seafood cannot be assured to consumers unless there is adequate quality control and quality inspection from the point when the fish is captured to the point when it is sold to the consumer.

Faced with this apparently extreme position, one might ask, If other foods in supermarkets do not require mandatory inspection, why is inspection of seafood necessary? Whereas meats and poultry are held under controlled conditions from the time they are in the slaughter-house to the time they are delivered to the retail stores, fish may at times remain in a fishing vessel or in a plant for seven or more days under conditions that are not regulated and often not controlled. Thus, while meats and poultry reach the supermarket with a reserve of high quality, seafood may reach it at the threshold of quality that is between acceptable and unacceptable.

Grade Labeling

In the United States, the USDC Inspection Service differentiates among four grades: A, B, C and Substandard. U.S. Grade A means that the product possesses good flavor and odor characteristic of the species, and all other required characteristics of size, shape, color, freedom from detects, and so on. U.S. Grade B means that the product possesses reasonably good flavor and odor characteristic of the species. Also, it may rate as highly in appearance, size, and so on. U.S. Grade C means that the product possesses minimal acceptable flavor and odor characteristic of the species, but is still free of offensive odors and flavors. Below these quality characteristics, the product is classified as unacceptable and its useful shelf life has ended.

For the quality assurance program, only U.S. Grade A products are accepted. But it can readily be seen that *grade labeling*—specifically U.S. Grade A—and *quality assurance* are not synonymous terms. One of the differences between the U.S. Grade A label and the assured high-quality logo identifying the quality assurance program is that while they both mean that the product was of high quality when it was inspected, only the logo guarantees that the quality will be high when

Official Grade of Seafood.

the product is consumed. The reason for this difference is that under the customary USDC inspection program, a product that qualifies for the U.S. Grade A label at the time of inspection might deteriorate in quality during distribution or some time before consumption. A second difference is that the program of assured quality guarantees only the organoleptic quality of the product; it does not guarantee such other aspects as workmanship, safety, required specifications, and so on, but rather relies on the use of the USDC Inspection Service to cover them. Therefore, it is best when a seafood product carries both the U.S. Grade A label and an assured quality logo.

Planning and Program Coordination

In order to assure the quality of seafood to the consumer, there is a need for much intensive and detailed knowledge, for much planning of activities in advance of execution, and for implementation of planning in full coordination with all participants in the program. We will give more attention to these separate aspects of the program in chapter 11. Here we want to emphasize the last factor, which we consider to be essential to the success of any program of quality assurance: the insistence that it starts with products of such high reserve quality that they last up to the moment of consumption.

High Initial Reserve Quality

For seafood, as for any other commodity, the level of quality that counts is that which is determined at the time the commodity is put to its intended use. Therefore, we can firmly state the following:

It is not enough that fish are of high quality when they are landed in port.
It is not enough that seafood is of high quality when it leaves the processing plant.
It is not enough that seafood is of high quality when it is at the retail counter.
It is not enough that seafood is of high quality when it is purchased by the consumer.
It is when seafood is consumed that its level of quality must be high.

And since it is only when seafood is consumed that its quality really counts, it must have enough reserve quality to last until that moment.

THE EFFECTIVENESS OF QUALITY ASSURANCE

It was stated earlier that seafood served in restaurants and that available in specialty seafood markets is generally of high quality, while seafood available at supermarkets is generally of poor quality. The discriminating difference between these conditions appears to be the presence or absence of *quality assurance*. Restaurateurs and operators of specialty seafood shops have a good knowledge of seafood quality. They have recognized the demand by their customers for high-quality seafood entrees and have reacted to this demand by insisting that the seafood processors deliver only products of high quality. The suppliers, in turn, demand the same from fishermen or middlemen. And, thus, the entire system is pressured into a combined effort to assure quality at the endpoint. Such a system implies the existence of standards of quality, certainly quality control, and a system of inspection and quality evaluation. But it is the entire chain that is deeply involved in assuring a satisfactory end result.

Quality assurance has been proven to work in a pilot project; it also has the potential to succeed nationwide in a very satisfactory fashion. The passage of the Fishery Conservation and Management Act of 1976 extended the jurisdictional boundary of the United States to 200 miles from its shores, encompassing many rich fishing areas and making it possible for it to recapture its position as the largest seafood producer in the world. The Gloucester Laboratory has amassed a reliable body of evidence to show that when the quality of fish fillets is maintained to meet the criteria of U.S. Grade A throughout their handling up to the point of sale to consumers, the demand for these products grows at very high rates.

The details of the economic impact of quality assurance are given in part 3. Here we want to look at the data found in figure 13.1 because it highlights the core of the results of that program. The effectiveness of the quality assurance of seafood is reflected in the sustained rate of growth of consumer sales, which not only grew, but grew at an exponential rate. The facts illustrated in the figure are especially impressive because they were generated mostly in the supermarket arena, exactly where the quality of seafood is not generally assured and sales are rather stagnant. There, spurts of increased sales may at times result from advertising campaigns and other sales strategies, but they are often short-lived and have no impact either on the overall rate of sales or on per capita seafood consumption. The only sales strategy that has shown a sustained effect on the demand for seafood is quality assurance. The pilot project to which figure 13.1 relates started in two supermarkets in northeastern Massachusetts with product supplied by one processor. Within four years, the demand grew to such an extent that eleven processors and more than 1100 supermarkets dispersed over at least fifteen states were attracted to the program. And all of this growth occurred because of an unsolicited consumer demand that was generated only by the reputation of fish fillets of assured quality. There were no advertising campaigns, no special sales inducements, or other marketing strategies. Only quality assurance was employed.

Supporting the idea that quality assurance is the major ingredient needed to sustain a strong consumer demand for seafood is the growth in the volume of seafood entrees in restaurants, which outstrips that of any other entree item. In addition, the nationwide popularity of a relatively new item, fish sandwiches, has been propelled in such a way that they now rank as high or higher than other popular sandwiches. This growth is attributed to the very high quality of the seafood reaching the consumer via fast-food chains.

It is, of course, true that advertising and the diffusion of general information concerning the health and dietary aspects of seafood have helped to propel consumer demand forward, but would there have been repeated orders if the quality had not been high?

REFERENCES

Anon. 1949. *The distinguishing features of fish.* Fish Mongers Hall, London, EC4.

Anon. 1961. Frozen fried fish sticks. *Consumer Reports* 26:80–83.

Anon. 1965. Frozen breaded fish portions. *Consumer Reports* 30:235–237.

Anon. 1979. Major NZ study finds that . . . Higher quality is key to success in big Japanese market. *World Fishing* 28(12):5–6.

Anon. 1980. *Quality Excellence in the 80's.* Government of Canada, Fisheries and Oceans. Ottawa, Publ. No. W.E. P. 80/002E.

Anon. 1981. GAO asks for study of U.S. seafood quality. *Food Engineering* 53(2):31.

Dagbjartsson, B. 1980. Fish quality programs in Iceland. *Sou'wester (Supplement)* 12(13):6–7. (Reported also in *Fisheries Council of Canada Bulletin,* June 1980, p. 2).

Fitzgerald, R. 1980. Fresh fish. Amen. *Alaska Fishermen's Journal* 3(8):20–23.

Gorga, C., J. D. Kaylor, J. H. Carver, J. H. Mendelsohn, and L. J. Ronsivalli. 1979. The Economic feasibility of assuring U.S. Grade A quality of fresh seafoods to the consumer. *Mar. Fish. Rev.* 41(7):20–27.

Gorga, C., and L. J. Ronsivalli. 1982. International awareness for quality seafoods. *Mar. Fish. Rev.* 44(2):11–16.

Gorga, C., and L. J. Ronsivalli. 1983. Quality control and quality assurance — getting the difference straight. *Infofish Marketing Digest* (4)32–34.

Ishkov, A. A. 1975. Fish swim toward the table. *Golos Rodiny* (Voice of the Homeland) (52) Trans. Seattle, WA: Office of the Vice President for Research, University of Washington.

Laslo, E. (Editor). 1972. *The Relevance of General Systems Theory.* New York, NY: Brazillier.

Lie, O. 1980. Fish quality programs in Norway. *Sou'wester (Supplement)* 12(13): 7–8.

Machiaverna, A. 1977. Grade A labels boost fish sales by 20%. *Supermarketing* 32(7):1.

McLean, D. A. 1980. Quality enhancement an essential key to market growth home, abroad. *Sou'wester (Supplement)* 12(13):5–10.

Miklos, P. Undated. *Consumer Attitudes Toward Seafood: A Qualitative Research Report.* New York, NY: Miklos Research Associates, Inc.

National Research Council (U.S.). 1985. *An Evaluation of the Role of Microbiological Criteria for Foods and Food Ingredients.* Washington, D.C.: National Academy Press.

Pappenheimer, J. 1980. A look at Norway's fish handling regs. *Alaska Fisherman's Journal* 3(8):52–54.

Phillips, E. 1980. Fisheries Council of Canada Annual Conference. In *Eurofish Report: The Fishing Scene.* Turnbridge Wells, England: Agra Europe, Ltd., pp. FS/1 to FS/5.

Ronsivalli, L. J., J. D. Kaylor, P. J. McKay, and C. Gorga. 1981. The impact of the assurance of high quality of seafoods at point of sale. *Mar. Fish. Rev.* 43(2):22–24.

Staples, D. 1980. How do Japanese get their fish quality? *Commercial Fisheries News* 8(3):20–21.

Toffler, A. 1980. *The Third Wave*. New York, NY: William Morrow and Company, Inc.

Watkin, G. R. 1980. Centuries of quality control. *National Fisherman* 61(6):58–59.

Watson, L. R. 1979. Supermarketing fish: Fortunes favour the bold. *Australian Fisheries* 38(10):37–42.

Wells, W. E. 1980. Newfoundland fisheries overview. *Sou'wester* 12(8):2.

8

The Roles of
the Fisherman

The quality assurance of seafood is a process that starts at sea, therefore the first individuals who take part in it are the fishermen. They have a two-fold responsibility: first, to use harvesting methods that obtain fish of the highest possible quality; second, to minimize any quality loss while the product is in their care.

HARVESTING METHODS

When fish are harvested live in traps or by some methods of line fishing or seines, the quality is at its best. (One of the advantages of aquaculture is that all fish enter the handling chain in a live state, thus consistently assuring the highest initial quality.) Even if the fish are not taken on board while they are still alive—the best possible method—their quality will not deteriorate significantly, provided that the fish are taken on board soon after they expire. The method of catch is not a significant factor by itself. In fact, excellent quality can be obtained no matter which conventional method is used: long line, otter trawl, seine, and even gill net. For instance, gill nets are often implicated in the landing of fish that are of inferior quality; but upon consideration, one discovers that the problem is not with the type of net but with the timing between the set and the haul. Whenever the nets are left unattended for long periods, the trapped fish die from a lack of oxygen and begin to deteriorate in quality—a process that starts soon after death. Regardless of the fishing method used, therefore, fish must be hauled out of the water as soon as possible—especially in the summer months, because the rate of deterioration is more rapid at the higher temperatures of the sea water.

QUALITY CONTROL METHODS

The second part of the fishermen's responsibility is to minimize the loss in the quality of the harvested fish during the time that the fish remain aboard the vessel. Theoretically, the optimum method is to process and freeze fish immediately after catching and to store them in a freezer, set at about −17.8° C (0° F) or below. In this way the quality of the product is retained at a high level for months; freezing and storing at even lower temperatures extends the shelf life for longer periods. (Since most U.S. fishing vessels do not have facilities to freeze fish, this section will deal mainly with the handling of unfrozen fish at sea.)

Once the fish are taken out of the water, fishermen can control both the temperature at which they are held and the time it takes to bring them to port. This, then, is the objective that fishermen should pursue: They should use the best handling procedures and should bring to port only fish of the highest possible quality.

Described below are some methods that are used and some that may be considered for use at sea to assure the quality of the catch: icing, chilled sea water, refrigerated sea water, superchilling, freezing, and salting.

Icing

Fish harvested by most U.S. fishing vessels are cooled with ice that is put on board when the vessel leaves port, and this is the most popular method used to preserve the quality of the catch. (U.S. vessels that fish for tuna are exceptions because they generally fish for long periods and generally travel to distant fishing grounds; therefore these vessels do have and employ mechanical refrigeration to freeze their catches.)

In the summer, the amounts of ice taken on board must be higher than in the winter. In fact, ambient temperatures in the winter may be so cold in the northern fishing grounds that, on rare occasions, no ice may be needed. The amount of ice needed is also a function of the length of the trip and the size of the catch. Since many U.S. fishing vessels are relatively small, and the fishing grounds to which they go relatively close to shore, many fishing trips are of about one day's duration. Such boats are generally called *day boats* and do not take much ice with them. Most sportfishing boats also go out for about one day or less, and since some of the fish caught by sportfishermen finds its way into commercial distribution systems, this discussion applies to them as well.

Icing is accomplished in several ways. In the earliest days of its use, fish were cooled by a pile of ice that was placed near the pile of fish— a very ineffective use of ice for cooling. Later, fish were cooled in pens (compartments) by layering them with ice. While this was an improvement over the early method and in some cases is still used, it does present disadvantages. One of the problems is that the pens can be several feet high. In this case, the bottommost fish are subjected to considerable pressure, particularly when ice, in lump form, exerts point loads on them under very high pressure. In addition, the bottommost fish are subjected to a contaminated bath of water that forms from the melting ice and carries with it the surface bacteria, slime, and expressed juices from the fish over which it washes.

Another problem with ice cooling in pens is that too often the layers of fish outnumber the layers of ice, a situation that does not take complete advantage of the cooling effectiveness of ice. In order to improve on this technique, most pens may be fitted with shelves that have the effect of substantially reducing the weight on the bottommost fish, but this improvement does not solve all of the problems.

In some cases, boxes are used instead of the pens. The boxes hold about 91 kilograms (200 pounds) of fish and enough flaked or crushed ice to adequately surround the fish. This technique further protects the quality of the fish because pressure and contamination are reduced; and the use of flaked or crushed ice increases the area of contact between fish and ice to create a more effective cooling system. The use of boxes also facilitates unloading of the vessel. In effect, the use of boxes is a reasonably satisfactory practice and has become more popular in recent times. Whether fish are stowed in boxes or in shelved pens, it is most important that enough ice be used, that the layers of fish be no greater than one fish thick, that the ice be in the form of small particles (crushed or flaked), and that vessels make trips of no greater duration than a few days.

Each fish should be stowed in such a way that it is completely surrounded with ice because this is the quickest way to lower the temperature of the fish: The transfer of heat from a warm surface to a cold one is most rapid when the difference between the temperatures of the two surfaces is greatest. When any part of the fish surface touches another fish, the rate of heat conduction is slow, if not nil, because there is little or no difference in temperature between the surfaces. This is the reason layers of fish should never be more than one fish thick when fish are stowed in ice. Because ice has the unique ability to absorb heat without a temperature rise as it is converted from ice to water, however, the heat removal from the surface of fish in contact with ice remains at a satisfactorily high rate until the fish are cooled substantially. The latent heat of fusion, that is, the amount of heat that ice will absorb without a rise in temperature as it melts to its liquid state, is 79.9 kilocalories per kilogram of ice (144 Btu per pound of ice).

Calculations made by scientists at the Gloucester Laboratory using these and other data (including the moisture content of fish) suggest that fishermen should take an amount of ice equal to one-quarter to one-half the weight of the expected catch, depending on ambient temperature. Of course, one needs to quickly acknowledge the difficulty in estimating the size of the expected catch; but that is, in fact, what must be done. Ultimately, since the precise amount of ice needed depends on many other factors (for instance, the ambient temperature, the duration of the trip, and the heat conductivity of vessel walls), to do an effective job fishermen should overestimate the amount, until they gain enough experience to make more accurate estimates.

Because the spoilage of fish starts soon after they expire, and because the rate of spoilage is accelerated at high temperatures and slowed down at low temperatures, the sooner the temperature of the fish can be lowered, the better. Consequently, the shelf life of fish that are cooled immediately is longer than that of fish that are allowed to remain at warm ambient temperatures for any significant length of time. As mentioned earlier, the size of the ice particles should be as small as possible in order to sustain a high cooling rate. Finally, we do not recommend that the ice contain bacteriostats or bactericides, a practice that has been, and may still be, employed to some degree. The ice must be from potable water (safe for drinking) in order to prevent contamination of the fish.

Chilled Seawater

The use of chilled seawater (CSW) to cool the catch and to keep it cool is uniquely suited to certain applications, such as in commercial fishing for squid. This method of cooling requires the use of watertight tanks or compartments to hold the CSW and the fish. The CSW is made from ice that is loaded aboard the vessel just before it departs and seawater that is pumped in at sea as soon as possible to avoid clumping of the ice. The use of harbor water must be avoided, of course, because of its likely contamination. The recommended proportion of the components in CSW is about one part water and two parts ice; the proportion of CSW to fish is about three parts CSW to four parts fish. Thus, for every 4000 kilograms of fish expected to be caught, the CSW formula would be 1000 kilograms of seawater plus 2000 kilograms of ice. In English units, for every 8000 pounds of fish expected to be caught, the CSW formula would be 2000 pounds of seawater plus 4000 pounds of ice.

CSW has a number of practical advantages over ice. From the standpoint of cooling effect, CSW is more effective than ice in any form. Just as flaked or crushed ice is an improvement over ice lumps for cooling fish because of the greater surface contact between fish and ice, so CSW is an improvement over flaked or crushed ice for the same reason. CSW makes possible maximum contact because the molecules of cold water, due to their much smaller size and greater mobility than the smallest ice particle, can make contact with points on the fish that are inaccessible to the smallest ice particles. Because of the much greater surface contact between fish and coolant when CSW is used, the rate of cooling is much greater than with ice alone. The water component of the CSW removes heat from fish; it also removes other heat entering the system through the container walls; and the heat in the water is gained by the ice, causing it to melt. The temperature of the CSW will remain constant as long as any ice remains in the system. Because of the presence of salts from the sea water and other soluble substances from the fish, the freezing point of the CSW will be somewhat lower than 0° C (32° F). It should be noted that because of the fluidity of CSW, the fish are cooled not only more rapidly but more uniformly, so that there will be no warm spots in the system, as is the case when ice alone is used.

Another important advantage of CSW is the buoyancy effect it has on the fish. Since the specific gravity of fish is approximately the same as that of water, the fish in a CSW medium are in a weightless relation to one another, so that there is no extreme weight effect on the bottommost fish. Accordingly, even the most delicately structured species can be satisfactorily held in CSW. Even though the bottommost fish are under a higher hydrostatic pressure than the uppermost ones, they are not subjected to any significant adverse pressure effects because of the equal distribution of the pressure around them. This was a major consideration for fishermen who had investigated the desirability of converting their fishing effort to the harvesting of squid. Squid have a delicate physical structure and, unless they are carefully handled, readily undergo damage during conventional stowage in ice.

Still another advantage of CSW is the ease with which fish suspended in it can be unloaded from the vessel. With increased pressure by regulatory agencies to eliminate the use of pitchforks for handling and unloading of the catch (a practice of long standing), one of the alternatives introduced is the use of a large mechanical pump that applies the vacuum principle to unload large quantities of fish from the vessel in a relatively short time. CSW, because of its fluid nature, lends itself very well to unloading the catch suspended in it by a mechanical pump (figure 8.1).

Finally, another advantage to the use of CSW is the effectiveness of its cooling. Assuming that the amount of ice that has been added is adequate, because of the total contact between fish and CSW and the consistently low temperature of the CSW, the temperature of fish chilled with CSW will be lowered more rapidly and to a lower degree than that of fish chilled with ice alone.

In at least one application, the advantage of CSW over ice depends on the end use of the product. One of the effects of holding scaly fish in CSW is that they lose some of their scales as a result of the abrading action that occurs under the constant motion of the system. The amount of scales lost depends on the species and on the degree of agitation of the system. The loss of scales may be advantageous or not, depending on whether the fish is going to be scaled anyway and whether it is desirable or necessary to leave the scales on.

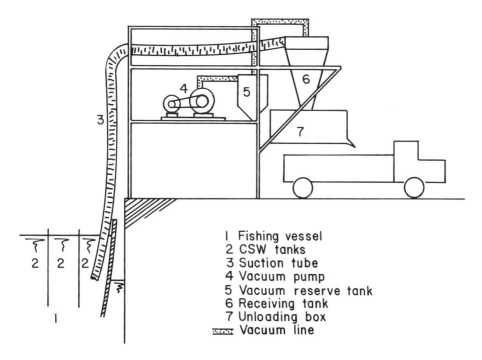

1 Fishing vessel
2 CSW tanks
3 Suction tube
4 Vacuum pump
5 Vacuum reserve tank
6 Receiving tank
7 Unloading box
⚏⚏⚏ Vacuum line

Figure 8.1. A Vacuum Pump for Unloading Fish.

In terms of vessel stability, CSW has some disadvantages because it is a fluid system and, as such, may subject the vessel to extreme and dangerous load shifting, particularly in heavy seas. To mitigate this risk, large compartments must be fitted with baffles or internal walls (figure 8.2), which serve to prevent large surges. Another means to mitigate the hazard of load shifting is to use a number of small containers that can be held securely in the hold (also, small containers are transportable). The cost of these modifications should be considered, but their economic feasibility has already been demonstrated in the conversion of the vessels used to fish for squid.

One possible disadvantage of CSW is the potential for undesirable chemical and physical changes. There have been reports that fish in CSW tend to absorb salt, that they may become rancid, and that they are apt to lose some color. Learson and Ampola (1977) reported on these and other reports that adverse effects are not so severe as to cause concern, except when the fish are left in the CSW for unduly long periods and/or the temperature of the CSW is permitted to rise due to an insufficient amount of ice used in the preparation. In work conducted by Learson and Ampola (1977) at the Gloucester Laboratory, none of the adverse reactions were noted. In these experiments, the CSW temperature was never allowed to rise, and the fish were kept in the CSW for no more than a few days. If these were to become real problems, however, as they might under extended stowage periods,

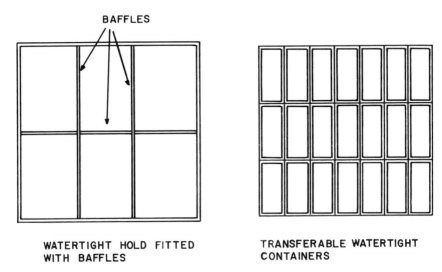

WATERTIGHT HOLD FITTED WITH BAFFLES   TRANSFERABLE WATERTIGHT CONTAINERS

Figure 8.2. The Use of Baffles or Containers to Prevent the Shifting of Fluid Loads in Fishing Vessels.

one could protect each fish by putting it in a plastic casing, much as is done with frankfurters.

One of the effects of chilling fish with CSW is the effect on the composition and development of the microflora. A theoretical effect, which has not been adequately investigated, is that CSW should control the size of the population of the aerobic bacteria inhabiting the system below the CSW surface (these bacteria are normally responsible for most of the loss in the quality of seafood). The reason for this effect is that the CSW sharply curtails the oxygen availability to the bacteria. This and other effects, such as those of lower temperature than is attainable in iced storage, are worthy of a more detailed investigation.

Near the beginning of this discussion, we stated that CSW should contain two parts ice and one part sea water, and that this proportion could be used to chill four parts of fish to a temperature of 0° C (32° F) or lower and maintain it at that temperature until port is reached. But these proportions are recommended as a first approximation, and it is expected that adjustments may have to be made for each vessel. Some of the variables on which these proportions depend are difficult to assess. While the most elusive is the prediction of the amount of expected catch, another important factor is a determination of the heat gain. This determination involves an equation that includes data on the thermal conductivity of the CSW compartment wall, its dimensions, the temperature differential, and so on. For example, in one CSW experiment conducted by scientists from the Gloucester Laboratory, the CSW compartment walls were well insulated, and this single modification reduced the value of the heat gain to such a level as to make a dramatic reduction in the ice requirement possible. The proportions of water, ice, and fish that were eventually determined to be satisfactory in this particular case were 2:1:7. Comparing this ratio against 1:2:4 recommended for uninsulated vessels, it can be seen that the cost of ice per unit of catch was reduced by about 70 percent when the system was insulated.

Other Cooling Methods

Three other methods for preserving fish by cooling have been used aboard many foreign and a few U.S. fishing vessels: refrigerated sea water (RSW), superchilling, and freezing. All of these methods require mechanical refrigeration. The NMFS has devoted decades of research to methods of preservation by cooling, and its laboratories at Gloucester, Massachusetts, and Seattle, Washington are a reservoir of information on all of these methods.

Refrigerated Sea Water (RSW). In the application of a refrigerated sea water (RSW) system for cooling the catch, the hold of the vessel is modified, as it is for using chilled sea water (CSW), but separate, smaller containers are not used with RSW. The major difference between the two systems is that the RSW is cooled by mechanical refrigeration, whereas the CSW is cooled by ice. Accordingly, the use of RSW requires the installation of a refrigeration system with such ancillary equipment as pumps, filters, and so on (figure 8.3). RSW has two advantages over CSW:

1. There need be no concern that not enough ice or too much ice is taken on any fishing trip (either case represents unavoidable but excessive costs).
2. The RSW temperature can be controlled to any level, down into the freezing range.

It should be noted that lowering the temperature below −2° C (28.4° F) requires the addition of salt or another solute (dissolvable substance) because sea water freezes at that temperature. RSW has been used to cool menhaden, sardines, salmon, and halibut.

Superchilling. Superchilling has been defined as any cooling process that lowers the temperature of seafood to a level within the narrow range −3 to −1° C (26.6° F to 30.2° F). In that temperature range, a small amount of the water in seafood will be frozen. The product will be somewhat rigid, but it can be deformed by the application of moderate pressure. At temperatures above that range, none of the water in the product will be frozen, while at temperatures below, there will be

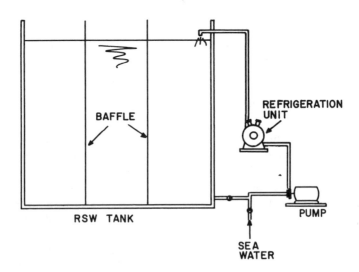

Figure 8.3. The Basic Elements of the RSW System.

enough water frozen within the product to make it quite rigid, and, in this state, the product cannot be deformed even under heavy pressure. Superchilling has also been labeled *supercooling, light freezing,* and *partial freezing.* It can be accomplished either by mechanical refrigeration, which is used to simply cool the compartments in which the product is kept, or by immersion of the product in RSW, which is brought to and maintained at temperatures in the superchilling range. Although this process has not been widely used, theory and the few reports on its application indicate that it results in a significant increase in product shelf life.

Freezing. Freezing of the product occurs when its temperature is lowered to levels below $-3°$ C (26.6° F). (Products are usually frozen to much lower temperatures.) As described above, a product is considered frozen when enough of its water has been solidified so that it will not be deformed even under heavy pressure. Freezing preserves seafood for long periods (months to years), depending on the holding temperature and the packaging used. When freezing is done properly (cold enough temperature and protection from dehydration, toughening, and rancidity), not only is the product preserved for a long time, but also its quality remains high. Freezing can be accomplished in the same fashion as superchilling, except that it involves much lower temperatures, and, in cases where fish are frozen in brine, the salt content of the brine must be raised in order to obtain an adequate depression of its freezing point (the lowering of the freezing point due to the addition of salt to the water). When the salt used is sodium chloride, and it usually is, the lowest brine temperature that can be attained is $-21.1°$ C ($-6°$ F) since that is the eutectic point of the salt. (The eutectic point of salt and other details concerning the freezing process will be covered in the next chapter.) To reach lower temperatures, the freezing point depressant to be used must have a lower eutectic point. Calcium chloride, which has a eutectic point of $-55°$ C ($-67°$ F), is one such substance.

As in the processes used to chill fish described above, the rate of freezing is faster when the product is brought into direct contact with the coolant. For instance, the process is faster when the product is frozen in brine than when it is frozen in freezer rooms in which a coolant cools the air to the same temperature as the brine and then the air cools the product. Fluid media other than brine have been used to freeze fish. These are classified as cryogenic liquids because they remain fluid at freezing temperatures.

Salting

Salting is a form of preservation in which a sufficient amount of salt is added to the fish to lower its water activity to the point that it inhibits the growth of spoilage bacteria. In general, lean fish are preferred for salting. The salting of fatty fish does not result in a satisfactory product for at least two reasons. The first is that salt penetration is impeded in flesh that contains fat, thus extending the salting time and subjecting the product to some quality deterioration. The second is that salt acts as a catalyst in the process that causes rancidity, and, whereas the off-flavors that develop in rancid lean fish are tolerable to a degree, those that develop in rancid fatty fish are less tolerable. Whether for lean or fatty fish, the rate of salt penetration increases as the temperature is increased; however, there is also an increased risk of bacterial spoilage at higher temperatures. For this reason, it is recommended that salting be done at chill temperatures, albeit at a slower rate. Pure salt is recommended over salt containing impurities such as magnesium, iron, and calcium, because despite some advantages such as improved color and firmer texture due to the presence of these elements, the disadvantages are more serious. They include retardation of salt penetration, bitter flavor, and acceleration of rancidity (Ronsivalli and Learson 1973).

Salting is classified in two ways: hard cure and light cure. To salt fish by the light-cure method, the fish are beheaded, eviscerated, and split ventrally. The backbones are removed and then the fish are washed to remove blood and any visceral remnants. They are then layered in large tubs, with layers of salt in between. The proportion of fish to salt is about 12:1 by weight. After about twenty-four hours, so much brine forms from the water expressed from the fish that the fish float in it. The fish are held submerged by weighted covers that fit inside the tubs. After two to three days they are removed from the tubs and laid in stacks to allow them to drain under their own pressure. The fish may then be dried in the sun or in a drying chamber upon reaching port. The salt content of lightly cured fish is 8 to 9 percent. If the light-cure method is chosen, the only change in gear is the addition of tubs or vats. As is the case for CSW, the size of the tubs should be such that they do not pose any risk to the stability of the vessel.

In the hard-cure method, the fish are prepared in the same way as above, but they are layered with the proportion of fish to salt at about 3:1. Also, the water leaving the fish is not allowed to form a brine but, instead, is permitted to drain off. Accordingly, tubs are not required but may be used provided that drainage is possible. The fish could be simply piled in stacks, but because of the motion of the vessel, the stacks would have to be confined by placing them in pens. The required salting period for this process runs for about two weeks; therefore, for most if not all fishing trips, the process would start at sea but would have to be completed on land. The stacks of fish, which are piled to a height of about four feet, are rearranged a few times during the process so that the fish at the bottom of the pile are relocated to the top position. In this way, the pressure to which they are exposed is equalized. The salt content of fish that have been salted by the hard-cure method is about 30 percent. The hard-cure method produces a product that is relatively dry, containing about 20 percent water. It will have a much longer shelf life than that produced by the light-cure method. However, the hard-cured product is eventually apt to become more rancid than the light-cured one.

When fishing trips are relatively short (two to three days), it is not necessary to consider salting any of the fish on board. But salting should be one of the processes to be considered for fish caught in the early part of a relatively long trip. While salted fish may not be in great demand in the United States, there is nonetheless a large market worldwide. Consequently, this preservation process has a reasonable market potential.

ADDITIONAL COMMENTS

Fair remuneration to fishermen is an important factor in the success of the quality assurance program. According to recent reports, not only are fishermen adopting most of the practices recommended here, but there is also widespread recognition that their efforts do deserve fair compensation (Wald, 1986).

REFERENCES

Kreuzer, R. (Editor). 1969. *Freezing and Irradiation of Fish*. London, England: Fishing News (Books) Ltd.

Learson, R. J., and V. G. Ampola. 1977. Care and maintenance of squid quality. *Mar. Fish. Rev.* 39(7):15–16.

Nickerson, J. T. R., and L. J. Ronsivalli. 1980. *Elementary Food Science*, 2nd ed. Westport, CT: AVI Publishing Co.

Ronsivalli, L. J., and D. W. Baker. 1981. Low temperature preservation of seafoods: A review. *Mar. Fish. Rev.* 43(4):1–15.

Ronsivalli, L. J., and R. J. Learson. 1973. Dehydration of fishery products. In *Food Dehydration, Vol. 2*, 2nd ed. W. B. Van Arsdel, M. J. Copley, and A. I. Morgan (Editors). Westport, CT: AVI Publishing Co.

Stansby, M. E. (Editor). 1976. *Industrial Fisheries Technology*. Huntington, NY: Robert E. Krieger Publishing Co. Inc.

Wald, M. L. 1986. Down East, the goal is fresher fish. *The New York Times*, April 16, p. C7.

CHAPTER

9

The Roles of
the Processor

Processors represent the second major element in the chain of activities that takes seafood from the sea to the table. The processor is anyone who does any work on fish or shellfish, taking it from the form landed by the fishermen and converting it to one demanded by the consumer: for example, fillets, steaks, shellfish meats. The processor may also produce such products as fish blocks, which are then sold to secondary processors who produce the consumer-type products: for example, breaded fish sticks, breaded fish cakes, and prepared frozen seafood dinners.

A processor who is part of a team established for the purpose of assuring the quality of seafood must achieve two objectives. The first is to procure only those products that have the necessary reserve quality to last up to the moment of consumption. The second is to hold to a minimum the extent of deterioration of quality that the product undergoes while it remains within the environs of the plant. Although seafood processors produce a variety of products, we are restricting our remarks to the production of fresh and frozen fish, fish fillets, and the products made from them; we hasten to reiterate our contention, however that the principles of handling these products have broader application, as handlers of other seafood products will readily recognize.

122

PROCUREMENT

The processor's tasks are not as simple as they might appear. In the procurement of fresh fish, the processor must obtain product of such high quality that it lasts up to the moment of consumption, and yet, the processor is unable to make this determination alone, no matter what tests are used. The reason for this potential dilemma is that at the beginning of the product's shelf life there is a period of about one week (assuming that the product is held at about 0° C or 32° F) during which it is difficult to pinpoint the quality level exactly. As a specific instance, there is no test that can determine whether the fish was caught two days or four days earlier. The reason that one cannot normally differentiate between the levels of quality of fish that are up to one week out of the water is that all such fish easily pass the strictest standards of quality, even though it has been shown that those fish that were caught early will not, in most cases, be of high quality by the time they are consumed.

It cannot be emphasized too strongly nor too often that the only level of quality of any seafood for which the entire industry is held accountable is one that is sensed at the time of consumption. All too often the assumption is made that fish of seemingly high quality at the dock or in the processing plant will be of high quality when the consumer eats it. Such an assumption can be, and too often is, invalid.

When procuring fish for the *fresh* fish market, determining the quality of fish at time of purchase is only a first step. The second step is to determine whether the fish has an adequate reserve quality. This reserve can be determined only from a knowledge of time and temperature of holding the fish since capture and the time and temperature of holding until consumption. Obtaining this information can be difficult under ordinary circumstances, but would not be in a coordinated program that assures quality, because this important information would be readily available to all participants. In general, while preparing products for sale in supermarkets, a processor should not attempt to process any fresh fish that are more than two days out of the water nor should he convert to frozen products any fresh fish that are more than seven days out of the water.

When procuring fish for the manufacture of *frozen* seafood items, the processor must be concerned only with the product quality as it is at the time of purchase. The processing, in this case, is not expected to lower the quality to a significant degree, nor should there be any loss of quality during subsequent handling as long as the product remains frozen until it is used. Only if a product is thawed for a day or more, is it necessary to determine the temperature history of the product and the temperature at which the fish will be held for any length of time in the thawed state.

MINIMIZING THE DEGRADATION OF QUALITY

The second task expected of the processor who is engaged in a quality assurance program is to minimize the extent of the degradation of quality that the product undergoes between procurement and shipment to the next element in the distribution chain. The techniques used for this purpose preserve the quality of seafood up to its natural limits. These techniques are rather standard and vary mainly as a consequence of the market to which the product is directed. Accordingly, we will distinguish between techniques for marketing fresh and frozen seafood.

PRESERVATION OF QUALITY FOR THE FRESH SEAFOOD MARKET

Deterioration of the quality of fresh seafood occurs as a function of the holding temperature, the time of holding, and the availability of the enzymes of spoilage microorganisms. The processor must therefore keep the product at as low a temperature as possible, minimize the time that the product remains in his plant, and operate under strict observance of the sanitary code. These are standard practices (two more advanced practices—the Friotube system and radiation ionization—are discussed below).

Maintenance of Temperature

Maintaining the temperature of seafood by land-based elements of the handling chain is facilitated by the greater number of temperature controlling options available to them as compared to those available to most American fishermen. Also, whereas the handling of fish aboard a vessel is generally done at ambient temperatures, processors in increasing numbers are controlling the temperature of the processing plant.

While the processing of meats and poultry is done in rooms in which the temperature is maintained at 4.4° C (40° F), the processing of seafood is carried out at temperatures that vary from plant to plant because, unlike meat and poultry plants, seafood plants do not operate under strict temperature regulations. Many seafood plants, particularly the smaller ones, process at ambient temperatures; there is a growing number of processors, however, who have imposed a voluntary control on their processing room temperatures. Some seafood processors maintain a processing temperature of 10° C (50° F). Others operate at 4.4° C (40° F), and at least one processor has operated below the latter temperature.

There are two concerns with the issue of the temperature at which seafood plants operate: One is the safety of the product, the other is the eating quality. From the point of view of public safety, a low temperature may not be necessary. There is a theoretical reason for this condition. It is related to differences between the microbial floras of meats and poultry on the one hand and seafood on the other. The microfloras generally contaminating meats and poultry (but not fish) consist largely of mesophiles, which include bacteria that cause diseases in man. Hence, the strict regulations that govern the processing of meats and poultry. The microfloras that normally contaminate seafood largely comprise psychrophillic bacteria, which are not ordinarily as dangerous as the mesophiles. From the point of view of eating quality, since the enzymes from psychrophiles react at faster speeds than do the enzymes from mesophiles for any given set of processing room temperatures, seafood should be processed at even lower temperatures than meats and poultry.

Minimizing the Processing Time

Since the shelf life of fresh seafood lasts but two weeks, provided the products are held at 0° C (32° F), and since the quality of seafood is at a high level for little more than one week, it can be seen that time is a commodity that cannot be wasted by any of the elements in the handling chain. The processor should not take much more than one day to process fresh fish for the supermarket trade. Time is crucial, and the processor must improvise, if necessary, to meet the schedule. If under unusual conditions the product is exposed to high temperatures, the need to speed things up is even more important.

Sanitary Handling

Sanitary handling procedures in processing plants is a subject that has been covered in many readily available texts and publications, and a number of these can be obtained without cost from the FDA, the USDA, and especially the NMFS. There are, however, three points that merit some discussion here. First, for practical purposes, sanitary handling consists, to a small degree, of education and, to a large degree, of common sense. For example, we know that the human body carries bacteria and that these bacteria readily transfer to clothing. Bacteria are also picked up by the hands from different parts of the body. Is it not only common sense to wash hands before handling food, to wear a clean work uniform, to dip hands in a bactericide, and/or to wear sanitary gloves before entering a processing line?

Second, there is no alternative to commonsense sanitary practices. Sophisticated sanitary practices such as the use of bactericides will not undo the damage done to the eating quality of seafood by any lack of observance of ordinary sanitary practices.

Third, sophisticated practices are limited in their effectiveness and some may even present problems—as they do when chemical agents are applied directly to seafood. Several experiments have shown that a plain water wash by spraying causes a reduction in the total number of bacteria by about 90 percent; on the other hand, investigations that have been undertaken to evaluate the sanitary effects of agents such as chlorine sprays, brine dips, and, earlier, antibiotic dips and antibiotic ice have not produced convincing evidence that these treatments are much better than washing with plain water and employing sanitary procedures. In addition, one of the problems arising from the availability of these antimicrobial agents is that they mistakenly may be relied upon to compensate for the procurement of unsanitarily handled fish or for unsanitary conditions within the plant. Also, potentially dangerous situations can result from the continued use of antibiotics. Notable is the development of resistant bacteria. This potential consequence has caused the United States Food and Drug Administration to ban the use of antibiotics for extending the shelf life of any food.

Needless to say, it is a completely different matter to use these agents—except antibiotics—to clean the plant, equipment, and utensils. The use of chlorine and other agents to clean the plant, cutting boards, and work areas, and as a sanitary dip for utensils has been found to be quite effective and is highly recommended.

The Friotube System

Keeping the processing room at cool temperatures is desirable from the standpoint of maintaining the quality of the products being processed, yet undesirable for worker morale and, to some degree, worker performance. To mitigate the discomfort of a chilling work environment, workers can be supplied with appropriate outer clothing to insulate them from excessive loss of body heat; in some cases, however, the added clothing may interfere with the ease with which workers would otherwise perform. Above all, the use of gloves impedes the ease with which a worker performs operations requiring dexterity of the hands and especially the fingers.

The adverse effect of cold temperatures on workers has accounted for a high turnover in one plant where an attempt was made to operate at below 4.4° C (40° F). In addition to the cost of high turnover rates, represented especially by the cost of training and the difficulty in attracting new employees, the higher energy costs to supply the refrigeration, the cost and maintenance of refrigeration equipment, and other ancillary costs all add up to an economic disincentive to refrigerate a plant to such low temperatures on a purely voluntary basis. For these reasons, a new system has been proposed (see Gorga 1983). This system, called the Friotube, confines the refrigeration only to the space that immediately surrounds the product.

Although a working prototype has not been developed, a design has been proposed that involves a system of refrigerated tubes or tunnels through which the products pass as they advance from station to station (figure 9.1). At the stations where the workers perform their required function, the temperature is kept at a more comfortable working level. The proposed system offers the advantage of adequately supplying the refrigeration needs to assure the products' quality at a much reduced cost over conventional methods, while at the same time maintaining a comfortable and efficient work force. It is expected that, once adopted by the seafood industry, the Friotube system will also find applications in the processing of other foods and in areas outside the food industry.

Ionizing Radiation

Whereas the radiation preservation of fruits, vegetables, and grains has been approved by the FDA, approval to preserve seafood remains in question. A description of this process may, at any rate, be appropriate.

It has been known for many years that foods can be adequately preserved by exposing them to ionizing radiation, a method that destroys bacteria and other microorganisms. The preservation effects of radiation on foods are similar to those of the heat process. Heat under pressure (e.g., 20 minutes at 1 kg/cm^2 or 15 lb/in^2) as well as high levels of radiation (e.g., 4.5 megarads) are used to sterilize foods. Lower levels of heat (e.g., less than 100° C or 212° F for a short period of time, namely, seconds to minutes) and lower levels of radiation (e.g., 300 kilorads) are used to pasteurize foods. Some food products, such as milk, would be so altered in their organoleptic quality if sterilizing levels of heat were applied that they are generally pasteurized. To a degree, this process preserves the product by destroying the bulk of spoilage bacteria and essentially all disease-causing bacteria.

Foods are sterilized when the applied level of energy is high enough to cause a 12-D destruction in *Clostridium botulinum*, the bacterium responsible for botulism, a dreaded food-borne disease. A 12-D destruction is said to be obtained when the number of botulinum bacteria is reduced by 99.9999999999 percent. By the application of a 12-D treatment, whether with heat or with radiation, the food products become

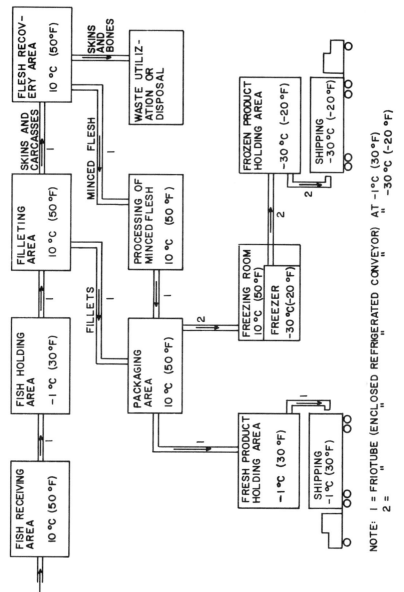

Figure 9.1. The Friotube Method for Controlling Product Temperature.

shelf stable and will remain preserved for years without the need for refrigeration or other preservation measures. It should be noted that products to be sterilized by radiation must receive a mild heat treatment to inactivate innate enzymes, because these have a high resistance to radiation and would cause the deterioration of the product through enzymolysis (the breakdown of tissue by enzymes). Products to be pasteurized by radiation generally do not require a pretreatment with heat, because during their relatively short shelf life enzymolysis is not a problem.

Foods are pasteurized, and retain their organoleptic characteristics, when the applied levels of energy serve only to destroy vegetative pathogenic microorganisms. The amount of heat applied in the pasteurization is sufficient to destroy disease-causing microorganisms, such as *Mycobacterium tuberculosis (hominis)*, the causative agent in human tuberculosis. Similarly, radiation can be applied at low levels causing a reduction of 99 to 99.9 percent in the number of contaminating bacteria, although, as mentioned earlier in this chapter, in the case of seafood, the concern is not with disease-causing bacteria, but rather with spoilage bacteria.

Pasteurized foods such as milk are not shelf stable and require refrigeration to obtain the possible short extension in shelf life: two to four weeks. Like milk, seafood cannot tolerate sterilizing treatments without adverse effects on quality, and when given a low radiation dose it also requires refrigeration and has a shelf life of about two to four weeks. Most experts agree that the preservation of seafood by radiation should be limited to low dose treatments (100 to 300 kilorads or 1 to 3 kilograys).

The rad is an acronym for radiation absorbed dose (one rad equals one unit):

1 kilorad   = 1000 rads
1 kilogray = 100 kilorads = 100,000 rads
1 megarad = 10 kilograys = 1000 kilorads = 1,000,000 rads

In general, when seafood is treated with about 150 kilorads (1.5 kilograys or 150,000 rads) of ionizing radiation, its quality is preserved to the extent that the shelf life is approximately doubled. The preservative effect is greatest when the treated seafood is held at the lower temperatures. It can, therefore, be seen that refrigeration is necessary as an adjunct to the process. Seafood treated in this way is handled as fresh, and there is no discernible organoleptic difference between irradiated and nonirradiated seafood.

The findings of one more study are relevant (Carver et al. 1969). The purpose was to investigate the effects of irradiating the fish as soon as brought aboard the vessel. It was determined that when the fish were packaged in oxygen-impermeable plastics, a dose of only 50 kilorads appeared to secure the same shelf life as a dose of 200 kilorads applied to fish landed ashore. From this study it was concluded that the sooner the fish is irradiated, the smaller the dose required to secure comparable extensions in shelf life. However, one also has to consider that the fish irradiated at sea were packaged in gas-impermeable plastic much earlier than fish treated on land. The question therefore arises, Is it only irradiation at the earliest possible moment that determines the required dose, or does the earliest possible elimination of oxygen also play a major role in the result? As far as we know, there are no data to shed light on the question. But it would appear that an investigation of this phenomenon is warranted.

Although it is outside the scope of this book to attempt a coverage of even the minimal relevant elements of the bacteriology of foods, food-borne diseases, radiation processing of seafood, nutrition, and the public health aspects of food processing, it is reasonable to mention radiation preservation because of its potential in the future preservation of seafood. It should be added that, at this writing, the radiation process is approved for only a few food products in the United States. The FDA has proposed a regulation permitting a general use of radiation at levels up to 1 megarad (1,000,000 rads), but whether or not approval is finally given will depend, to a large extent, on the response that the agency receives from the public, the industry, and other interested groups.

Other Forms of Preservation

Preservation of seafood can also be accomplished through salting, canning, smoking, and drying, among other methods. The selection is largely determined by the market to which the final products are directed. We will not cover these methods, however, because they have not been found to present technological problems in relation to the maintenance of quality, which is the primary concern in this book.

PRESERVATION OF QUALITY FOR THE FROZEN SEAFOOD MARKET

The greater acceptance of fresh seafood by the U.S. consumer and a higher retail value for fresh than for frozen seafood create an anomalous situation, as will be seen in more detail in part 3. While it commands a lower retail price, it costs more to produce frozen seafood than to produce fresh. However, frozen seafood objectively has more added value than fresh seafood of the same quality because it has a much longer shelf life; it does not need to be repackaged by the consumer for freezer storage in the home; and since it has already been frozen, energy does not need to be expended by the consumer to freeze it for storage.

Foods are frozen in order to preserve their quality over relatively long periods of time (months to years). The lower the temperature to which they are brought, the longer will be the time during which their quality will be preserved. It is believed that the freezing of foods to preserve them was practiced by ancient peoples who inhabited parts of the earth that at times experienced ambient temperatures well below the freezing point of water. The industrial freezing of foods was apparently introduced by Clarence Birdseye during the 1920s, when he developed a process for freezing foods in small packages suitable for retail merchandising. He showed that the quality of a number of foods, including fish, fruits, and vegetables, could be preserved for long periods by freezing them and holding them at freezing temperatures. Subsequent developments in freezing equipment, media, and techniques have led to the sophisticated frozen food industry of our time. It should be noted that much of the evolution in the freezing of foods has been in connection with the preservation of fishery products.

The rate at which foods are frozen has a bearing on their quality. If they are frozen slowly, there is an opportunity for some bacterial spoilage to develop because of the long time involved; there is also the opportunity for tissue damage due to the formation of large ice crystals. This second phenomenon occurs during slow freezing, when water molecules are able to migrate and to agglomerate, forming large ice crystals at sites where the first seed crystals formed. On the other hand, during rapid freezing cycles, the number of seed crystals is much larger and water molecules are unable to migrate to other seed crystals because they are frozen in their respective sites. Microscopic analyses of cross-sectional slices of fish tissue have proven the presence of fewer but larger ice crystals, and organoleptic and objective analyses of the product quality have proven that adverse quality changes are greater when the products are frozen slowly than when they are frozen rapidly.

Liquid Refrigeration Systems

In liquid refrigeration systems the product is immersed in a tank that contains the refrigerant, or is sprayed with liquid refrigerants, or is placed in a chamber in which the vapors from liquefied refrigerants act to freeze the seafood. These systems are used because freezing is most rapid when seafood is brought into direct contact with the refrigerant, when the refrigerant makes contact with the entire surface of the product, and when the refrigerant temperature is very cold. Refrigerants used in direct contact with food are known as *food freezants*. The most common food freezants are composed of various types of brine; others include liquid nitrogen, liquid carbon dioxide, and Freon-12.

Brine. Brine, a special type of liquid refrigerant, may be defined as an aqueous solution of one of various salts. Both the type of salt and its concentration may be varied to suit different applications. Generally, brines are made from ordinary table salt (sodium chloride) and water. The addition of salt to water takes advantage of one of the many properties of water; namely, the freezing point of water is lowered when substances are dissolved in it. Salt is a convenient, inexpensive, and relatively safe substance to use for this purpose. The higher the salt concentration, the lower will be the freezing point of the brine. This tendency exists until salt concentrations have reached the limit of 23.3 percent by weight. When the salt concentration is at this level, the freezing point of the brine will have been lowered to $-21.1°$ C ($-6°$ F). This is known as the *eutectic point* of sodium chloride, and it is the lowest temperature at which the brine can remain fluid. No matter how much higher the salt concentration is made, the freezing point cannot be lowered any further. In fact, as anomalous as it may seem, when the salt concentration is increased beyond its eutectic point, the higher the salt concentration, the higher will be the freezing point of the brine. As used in food science, the eutectic point is the point beyond which any increase in the concentration of dissolved salt ceases to further depress the freezing temperature of its solution.

Where sodium must be avoided or where it is necessary to depress the freezing point of water to levels lower than can be attained with sodium chloride brines, calcium chloride may be used. The eutectic point of calcium chloride is $-55°$ C ($-67°$ F). Calcium chloride brines, therefore, will remain liquid down to that temperature. As in the case of sodium chloride, the greater the amount of calcium chloride used to make the brine, the lower will be the freezing point of the brine. This trend holds until the amount of calcium chloride reaches 29.87 percent by weight. At that concentration of the salt, the freezing point of the brine is lowered to its eutectic point, and an increase in the salt concentration will reverse the trend and raise the freezing point. Despite the greater effectiveness of calcium chloride as a freezing point depressant for water, sodium chloride remains the salt of choice for the production of brines in the seafood industry.

Brines have been used by seafood processors for prechilling, freezing, firming tissues, salting, as a medium for canning, and as a base for pickling. Brines may produce adverse quality changes, however, if certain aspects of their use are not adequately considered. Weak brines, which are used as prechilling or firming dips, are often allowed to rise in temperature, in which cases they can no longer chill fish. Also, such brines may become diluted. Combined with the rise in temperature, weak brines result in the growth of bacteria that thereafter serve only to contaminate any products dipped in them. Strong brines have a tendency to corrode equipment, especially when their pH is lowered into the acid range. In applications such as salting and canning, it is preferable to use only the purest salt (less than 1 percent impurities). Some metallic contaminants, such as iron, may discolor the brine and may even affect the product. Other impurities, such as metallic sulfates, may impart a bitter flavor to the product.

Other Liquid Refrigerants. Other liquid refrigerants may be used to freeze seafood by direct contact; but they must meet with the approval of the FDA to insure that the wholesomeness of the product is not compromised. The approved liquid refrigerants currently in most common use are liquid nitrogen, which boils at $-195.8°$ C ($-320.4°$ F); liquid carbon dioxide, which boils at $-78.5°$ C ($-109.3°$ F); and Freon-12 (dichlorodifluoromethane), which boils at $-29.8°$ C ($-21.6°$ F). The continued use of Freon-12 and other halogenated hydrocarbons as refrigerants will depend on decisions regarding the damage to the earth's ozone layer that, as widely reported, is almost certain to occur when these compounds are released into the atmosphere.

Mechanical Refrigeration Systems

Freezing of food products is also accomplished with the use of mechanical refrigeration systems that employ cold air or cold contact plates.

The refrigerants used in both systems are classified into three groups. Group I refrigerants are those that are neither toxic nor flammable. They include liquid nitrogen, liquid or solid carbon dioxide, and some of the halogenated hydrocarbons. Group II refrigerants are either toxic or flammable or both; ammonia is the most commonly used. Group III refrigerants include those that are highly flammable and explosive, such as methane, ethane, propane, ethylene, and propylene. The latter group has only limited use, usually in systems that are already at risk of fire or explosion, in which case the use of a Group III refrigerant does not add significantly to the risk. A listing of some of the industrially important refrigerants in Groups I and II is shown in table 9.1. (Refrigerants belonging to Group III will not be discussed here.) It should be noted that the refrigerants are identified by a number preceded by the letter R. The identification numbers are set by, and

Table 9.1 Commonly Used Refrigerants

Identifying Number	Group	Chemical Formula	Name
R-11	I	CCl$_3$F	Trichloromonofluoromethane
R-12	I	CCl$_2$F$_2$	Dichlorodifluoromethane
R-22	I	CHClF$_2$	Monochlorodifluoromethane
R-500	I	CCl$_2$F$_2$ (73.8%) and	Dichlorodifluoromethane
		CH$_3$CHF$_2$ (26.2%)	Methyldifluoromethane
R-502	I	CHClF$_2$ (48.8%) and	Monochlorodifluoromethane
		CClF$_2$CF$_3$ (51.2%)	Trifluoromethyl-monochlorodifluoromethane
R-503	I	CHF$_3$ (40.1%) and	Trifluoromethane
		CClF$_3$ (59.9%)	Monochlorotrifluoromethane
R-504	I	CH$_2$F$_2$ (48.3%) and	Difluoromethane
		CClF$_2$CF$_3$ (51.7%)	Trifluoromethylmonochloro-difluoromethane
R-717	II	NH$_3$	Ammonia
R-728	I	N$_2$	Nitrogen
R-729	I	N$_2$/O$_2$/H$_2$O/CO$_2$/H$_2$/etc.	Air
R-744	I	CO$_2$	Carbon dioxide

Source: Ronsivalli and Baker 1981.

the only ones officially recognized by, ASHRAE (American Society of Heating, Refrigeration and Air Conditioning Engineers).

A simple mechanical refrigeration system consists of a room or chamber where the seafood is frozen, a refrigerant, an evaporator, a motor-driven compressor, a condenser, and an expansion valve (figure 9.2). The system is maintained at high efficiency by insulating the chamber to reduce the amount of heat gained from the outside.

AIR SYSTEMS

In systems that refrigerate with cold air, mechanical refrigeration is used to cool the air, which in turn cools the products. The air may

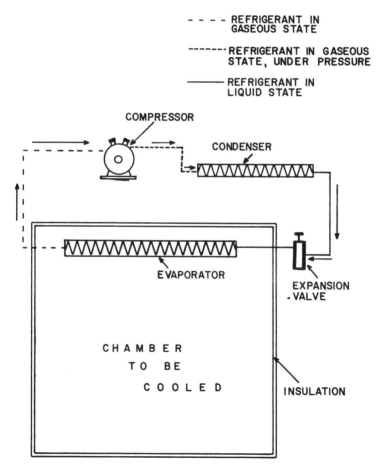

Figure 9.2. Basic Elements of a Mechanical Refrigeration System. (From Ronsivalli and Baker 1981.)

or may not be driven by fans. Although air, when stagnant, is a poor conductor of heat, it readily adds or removes heat from systems when it is put into directed motion or is allowed to circulate. Because of the effect of heat on its density (colder air is denser and heavier) and because of the ease with which it can be moved, as well as its unlimited free availability, air is a very convenient refrigerant.

Undriven Air. In many facilities, foods may be frozen and held at freezing temperatures with the use of a mechanical refrigeration system simply by lowering the temperature of the air in the freezer chamber to the desired level. It is this type of system that is used in domestic freezers and in some commercial facilities. Although the air is not driven by fans or by other means, it nevertheless does undergo motion as described below. Air that is not in motion cannot cool the product. In such applications, an evaporator (or system of expansion pipes) is usually situated near the ceiling of such holding rooms. The air that touches the evaporator is cooled, thereby becoming denser than the rest of the air in the room, and tends to descend to the bottom, where it displaces the warmer air. The warmer air rises because of its lower density and, upon reaching the upper part of the chamber, is cooled by the evaporator. The moving air is continually removing heat from the warmer surfaces it touches. When all the heat has been removed from the product, the air in the room will be cooled to a relatively stable temperature. A truly constant temperature cannot be attained, however, because the system continually gains heat through the walls, ceiling, and floor of the freezer chamber by conduction. Cold air is lost (and, of course, warm air is gained) when the door to the freezer is opened to put the product in or take it out. And, lastly, considerable heat is gained when the product is first brought into the freezer to be frozen. When the product is already cold before it is brought in, the warming effect is, of course, lower. Food placed in such a chamber is frozen by losing its heat to the falling cold air molecules, which, upon being heated, undergo expansion with a concomitant reduction in density by which they are buoyed upward again.

Should the food be without a protective wrapper, the cold air, which has lost its moisture to the even colder surface of the evaporator, will be under a low vapor pressure and will consequently act as a condenser for water vapor; the water vapor will leave the food because of the relatively high vapor pressure in foods. This is the cycle by which frozen foods dehydrate and is the reason food to be frozen must be protected either by an ice glaze or by a package impermeable to water vapor.

Slow-moving, undirected air molecules take a longer time to bring the product to a frozen state than do the driven air systems, which are discussed next.

Driven Air. The time for freezing foods in cold air is significantly shortened when the air is driven by fans. The high velocity of the cold air makes it possible for the surface of the food to be under a constant barrage of cold molecules, which facilitate the removal of heat because of their high number and their colder temperature as they come directly from the coils. From a quality standpoint, foods should be frozen in systems that use driven air. When the air is driven at high speeds, such systems are called *blast freezers* (figure 9.3). The air velocities in a blast freezer may be as high as 500 meters per minute (more than 1600 feet per minute).

Fluidized-bed freezers consist of an area in which cold air is driven upward through a perforated plate (see figure 9.4). Foods of relatively small size, such as shrimp, are directed over this bed of air and are buoyed up so that they are in constant motion, alternating between being buoyed upward and falling. The falling motion starts at the level that is reached when the upward momentum of the product is spent and the weight of the product particles is too great to be supported by the reduced pressure of the air column at that level. The upward motion of the particles starts soon after they fall to a level where the upward pressure of the driven air exceeds the downward pressure exerted by the particles. As they approach this point, the particles are slowed, then stopped, then accelerated upward. This rapid method of freezing is effective for preventing agglomeration of particles; thus, the production of individually quick-frozen (called IQF in the trade) shrimp is most efficiently and effectively accomplished by fluidized-bed freezing. This process has the potential for future application in the freezing of fish, fish fillets, and possibly other products.

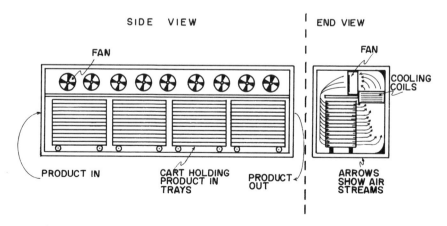

Figure 9.3. Tunnel Blast Freezer. (From Nickerson and Ronsivalli 1980.)

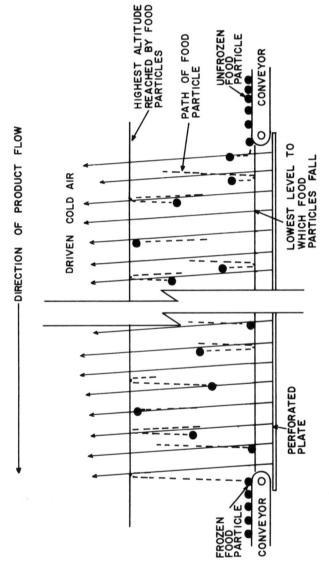

Figure 9.4. Movement of Particles in a Fluidized Bed Freezer.

138

Driven air systems are also used in *holding rooms*. Once foods are frozen to the desired temperatures, they are usually stored in a freezer in which small fans have the function of maintaining a minimum circulation of air.

CONTACT SYSTEMS

In some mechanical refrigeration systems, the removal of heat from the product is accomplished through contact with a refrigerated surface. Plate freezing is an example of such a system.

Plate Freezing. In plate freezing, layers of the packaged product are sandwiched between hollow metal plates through which the refrigerant flows (figure 9.5). The refrigerant is allowed to expand in the hollow of each plate, and then it is recycled in the mechanical refrigeration system of which it is a part. The plates, thus cooled to about −33° C (about −27.4° F) or below, are brought together, thereby assuring maximum surface contact with the product sandwiched between them. By this process, the product temperature can be lowered to about −17.8° C (0° F) in one and a half to four hours, depending on the thickness and starting temperature of the product. Although some plate freezers use a batch-type operation most operate as continuous systems. In batch-type operations, the plate freezer, using only one set of plates, is fully loaded and the freezing cycle is started. When the product is frozen to the desired temperature, the machine is unloaded of all of its products, and a new batch is loaded in the machine. Thus the product is frozen in batches. In a continuous operation, the machine, using as many as eight rows of plates, operates continually. Packaged products to be frozen are loaded between the plates in the first row. At the appropriate time interval, product from the first row of plates is automatically advanced to the second row of plates, while product from the second row of plates is automatically advanced to the third row, and so on. With each cyclic advance, the first row of plates is filled with unfrozen product, usually arriving by conveyor belt, and the last row is emptied of the product, which by this time has been completely frozen to the desired temperature.

Dehydrocooling

Dehydrocooling works as its name implies. It is a process in which the removal of heat is accomplished through the removal of water by evaporation (figure 9.6). This evaporation has a cooling effect, because the conversion of water to vapor requires the expenditure of energy by

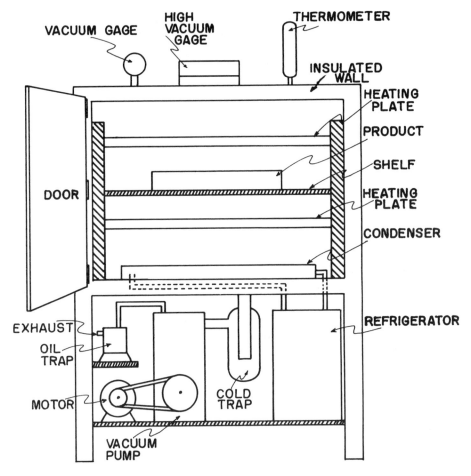

Figure 9.5. Batch-Type Plate Freezer. (From Nickerson and Ronsivalli 1980.)

the system at the rate of 539 calories for every gram of water vaporized (971 Btu for every pound of water vaporized). This value is called the *heat of vaporization*. What this means is that for every gram of water that is evaporated, 539 grams of water will be cooled by 1° C (for every pound of water evaporated, 971 pounds of water will be cooled by 1° F). Dehydrocooling is the principle by which a desert traveler could keep his drinking water somewhat cooler than the ambient temperature. A *desert bag* has a very slight permeability to water, and the evaporation of water from the wetted surface of the bag to the dry desert air causes the bag and its contents to lose heat. Dehydrocooling is also the principle by which a wetted finger held in the air allows one to detect from

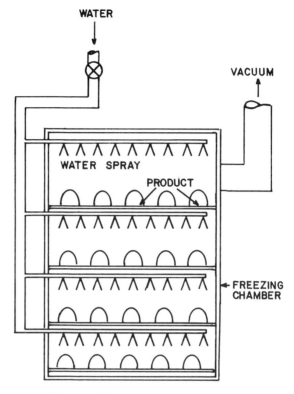

Figure 9.6. Basic Elements of a Dehydrocooling System. (From Carver 1975.)

which direction the wind is blowing, because the side of the finger facing the wind will feel cool due to the accelerated evaporation from the windward side.

Dehydrocooling has several applications. It is the principle employed when fans are used to cool us in the summer and is one of the most practical and efficient ways to cool leafy vegetables such as iceberg lettuce.

In experiments undertaken to cool fish by dehydrocooling, conducted at the Gloucester Laboratory, it was found that fish could be cooled from an initial temperature of 15° C (59° F) to 0° C (32° F) in approximately eighteen minutes (Carver 1975) (figure 9.7). (This is a remarkably fast cooling rate, and it leads us to believe that this line of investigation is worthy of continued study.) The effectiveness of the method, as with other systems, depends on a maximum surface exposure, which in this case can be achieved by placing fish in single layers on shelves made of wire mesh.

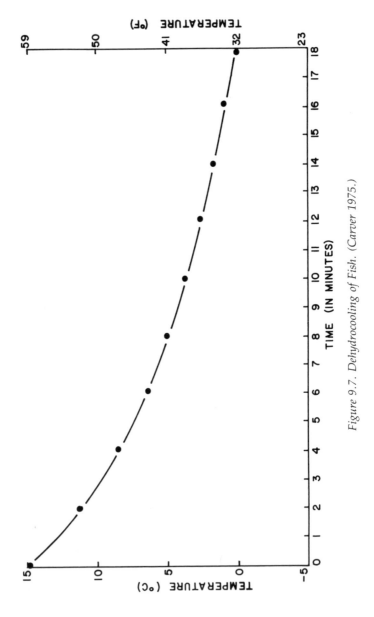

Figure 9.7. *Dehydrocooling of Fish. (Carver 1975.)*

Recognizing the undesirability of water loss by the product, a water sprayer was added to the dehydrocooler used in the experiments. Intermittent wetting of the product by a light water spray emitted by the sprayer easily replaced all the water lost by the product to evaporation, thereby eliminating any potential for a dehydration problem. One of the aspects of the process that still deserves to be determined is the cooling rate at temperatures below freezing.

High-Altitude Freezing

High-altitude freezing has been considered as a solution to the problem of food scarcity in developing countries with tropical or subtropical climates. Many of these countries are endowed with coastal areas where there is a relative abundance of edible marine resources; yet, the transfer of these resources to inland areas is made impossible by several circumstances. Generally, ambient temperatures are at such high levels that fish will spoil within one day and neither ice-making nor freezer facilities, nor insulated shipping containers are available in most, if not all, of the poor coastal areas.

In those areas where ice is available, the fish can be preserved for one week or more, and with a supply of ice some of the resource can be taken inland. However, many of the roads are poor and trucks have no mechanical refrigeration; therefore, only relatively short distances can be covered: about 320 kilometers (200 miles) before the seafood spoils.

High-altitude freezing takes advantage of the cold air (about $-50°$ C or $-58°$ F) that exists at altitudes normally used by jet aircraft: about 10 kilometers or 6.2 miles (figure 9.8). The principle is relatively simple. Fresh fish obtained at coastal locations are loaded onto modified aircraft in specially designed insulated boxes that allow the passage of cold air from the outside through forward intakes. During flight, the air passes through the containers and leaves through exhaust ports situated at the rear of the craft. It is calculated that the product will be frozen hard during flight in a period of three to four hours. Once frozen, the insulated boxes can be off-loaded at destination airfields or parachuted from the aircraft to areas without such facilities. The frozen fish will remain frozen in the insulated containers for adequately long periods to permit their distribution locally.

Valuable input to this solution, which was originated at the Gloucester Laboratory, was obtained from engineers at the Boeing Aircraft Company, where the designs for modifications to a Boeing 727 and for the insulated containers were produced (figures 9.9 and 9.10). In the design of the aircraft modification, attention was paid to its effect on the aerodynamics of the aircraft, the need to incorporate the insulated

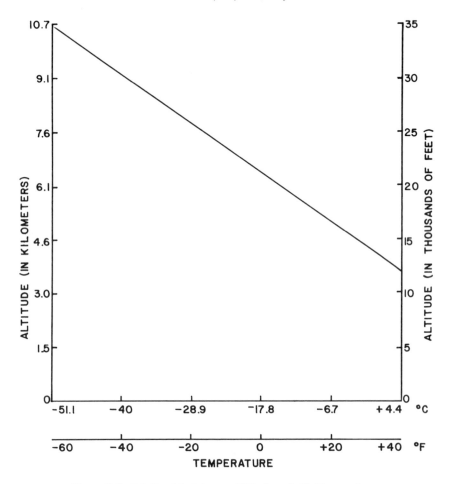

Figure 9.8. Relationship between Altitude and Air Temperature.

containers, placement of the product, the low density of the air at such altitudes, and the relative velocity of the air entering the forward intakes of the plane. High-altitude freezing appears to be theoretically sound. All of the apparent potential problems were considered and either discounted or resolved, but, of course, there is still the need to conduct pilot tests before its application can be considered.

The potential application of this method was envisioned as a short-term solution that might be financed by governments, by international agencies, such as the Food and Agriculture Organization of the United Nations, or by industries that might achieve economic gain from it: e.g., airline companies, manufacturers of containers, seafood shippers.

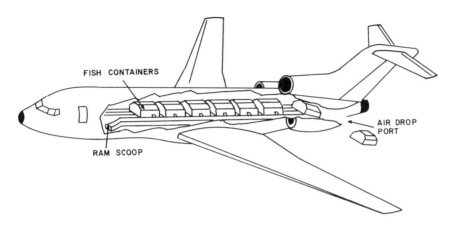

Figure 9.9. Modified Boeing 727 for Use in High-Altitude Freezing. (Courtesy Boeing Aircraft Corp.)

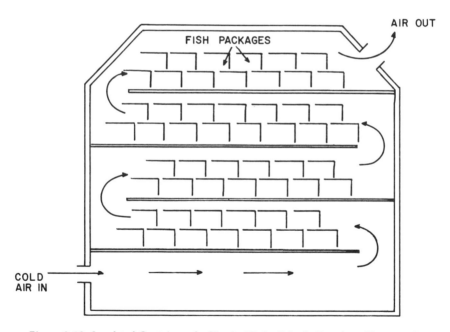

Figure 9.10. Insulated Containers for Use in High-Altitude Freezing. (Courtesy Boeing Aircraft Corp.)

USE OF FRIOTUBE

Although the Friotube system described earlier was designed for the processing of fresh fish, it has great potential for application in the production of frozen seafood for the same reasons given earlier: quality preservation, economics of processing, and worker morale.

THAWING OF FROZEN SEAFOOD

The need to thaw frozen seafood prior to reprocessing or preparation for consumption is an undesirable aspect of freezing as a method for preserving foods. Thawing—except when it is done with microwaves both commercially and in the home—is a time-consuming process that may result in partial loss of product quality. The surface of the product thaws first, and there is where the spoilage reactions occur. Surface spoilage is more of a problem when the ratio of volume to surface is large, as in whole large fish, or when, in the haste to achieve thawing quickly, the product is placed in a warm environment.

Products can be thawed satisfactorily in a refrigerator or refrigerated room set at about 2° C to 3° C (35.6° F to 37.4° F). With the length of time depending on the size of the product and other minor characteristics, this is too slow a procedure in industrial processing, and it requires too much space. Consequently, experiments were conducted cooperatively between the Gloucester Laboratory and the Raytheon Company (Learson, 1984) to prove the theoretical superiority of the large scale microwave thawing or tempering of frozen seafood blocks. Following the successful experiments, the industrial-scale thawing of foods by microwaves was introduced first to the seafood industry and then to the meat industry during the 1970s. The use of microwaves allows frozen seafood to be thawed or tempered (partially thawed) rapidly and without quality loss. Much of the industrial application of microwave energy is for tempering or partial thawing. (*Tempering*, as used here, means raising the temperature of the product so that it can be processed without completely thawing it.) Microwave energy, by its unique character, causes the temperature of the product to rise, although not entirely uniformly, in all parts of the product, internal as well as external. In conventional thawing procedures, on the other hand, thawing starts at the surface.

Microwave energy occurs as an electrical current in which the charge is alternated between positive and negative at rates of either 915 million or 2450 million times per second (both frequencies are currently authorized by the federal government for use with microwave ovens).

Since water molecules are polar (have positive and negative ends), they tend to twist as they are induced by the microwaves to align themselves first toward the positive and then the negative electrical charges. (All frozen food products contain some unfrozen water, and it is the unfrozen water molecules that react to the microwaves. For more details on the freezing of water in fish, see appendix 2.) The twisting action of the water molecules creates considerable friction within the system, and the resulting generation of heat raises the temperature of the system, thereby melting the ice and heating the product. The water produced from the melting ice becomes part of the heat-generating system. Therefore, the process is automatically accelerated. It should be noted that ice is not melted by the microwaves directly but by the heat of friction produced by the twisting molecules of already existing water.

In earlier applications of microwaves for thawing frozen foods, one of the problems encountered was uneven heating, which was due mostly to the erratic delivery of microwave energy—an innate shortcoming of microwave thawing. Because of the rapidity with which water is heated by microwave energy, and the lack of effect on ice, so-called *hot spots* developed readily and resulted in the partial cooking of the product before it was completely thawed. For many applications the development of hot spots is undesirable or intolerable, and therefore solutions to the problem had to be found. One was to apply the energy in intermittent bursts; during the interval between bursts, there is a dissipation of heat by conduction away from any hot spots. Another solution, applied mainly in some domestic units, incorporates a turntable; the food rotates and this motion compensates, to some degree, for the unevenness in the power delivery by changing the position of the food. The use of wave guides to reduce the unevenness in the power delivery is still another technique that has been used mainly in industrial units, but may be used in domestic units as well.

PACKAGING

The first element in the handling chain to package seafood is the processor. Although some fresh fish and meats may still be displayed and sold at retail stores without packaging, many are now packaged. An appreciation of the principles of sanitation and protection of the product quality would mandate the packaging of all of these products. Fresh fish are packaged mainly to protect them from invasion by microbes and other potentially harmful environmental contaminants. The package may also be used to inhibit the availability of oxygen because the rate of fish spoilage is slowed down in the absence of oxygen. There are many plastic films that prevent the entry of oxygen. These include laminated films containing a layer of aluminum foil, polyesters such as Mylar, and polyvinylidene chlorides such as Saran and Cryovac.

With one vital provision, the packaging used for fresh fish should be impermeable to gases. The provision is that impermeable packaging for fresh fish should be used only in systems where there is no chance whatsoever that the product will be exposed to temperatures above 4.4° C (40° F). The reason for this concern lies in the danger of an outgrowth of *Clostridium botulinum*, a genus of relatively ubiquitous bacteria. These bacteria grow in the absence of oxygen and produce a toxin that causes botulism, a potentially fatal disease. Since quality assurance requires a commitment to keeping the product at 0° C (32° F), there is no reservation in a quality assurance program about recommending the use of gas-impermeable packaging.

Because of the low temperatures at which frozen seafood is kept, the potential problem of botulism is of no concern for frozen products. Frozen fish are packaged to prevent not so much contamination as dehydration. Therefore, the package must be impermeable to water vapor. In addition, there should be no air space between the product and the package; otherwise the product will undergo dehydration at the site of any air space, regardless of the impermeability of the packaging material to water vapor. Furthermore, when the product is high in fat content, the package should also be impermeable to oxygen in order to prevent rancidity, a major deteriorative change in high-fat products. Lean fish may also develop rancidity, but the change is not as severe as in fatty fish. On the other hand, there is some evidence that in the absence of oxygen, some lean species containing a particular enzyme are apt to become tough during frozen storage.

In considering the properties that one should seek for packaging frozen fish products, therefore, there may be a dilemma in cases where the fish are of the type that contain the enzyme responsible for toughening. In these cases, one can select a package that is permeable to oxygen (it prevents toughening but cannot prevent rancidity) or a package that is impermeable to oxygen (it prevents rancidity but does not prevent toughening). In either case, the package must be impermeable to water vapor. It can be seen that there is no clear-cut recommendation for the packaging of these particular products: One can prevent, at least in part, either rancidity or toughening, but not both at the same time, when the product contains the enzyme responsible for toughening.

The situation is one that clearly deserves a continuing scientific investigation in order to be resolved. The chief facts to keep in mind are that rancidity that develops in lean species does not appear to be as severe nor to progress as rapidly as that in fatty species, and, in any case, its rate of development is slowed in direct relationship with the lowering of the temperature.

Insulated Shipping Containers

Shipment of seafood from the processing plant should be done in insulated containers, regardless of whether the product is frozen or fresh. Adequately chilled and frozen products should be protected from any possible temperature abuses while in transit, particularly if the shipment is to last for more than a few hours.

In order to illustrate the value of insulated shipping containers, we call attention to an experiment conducted at the Gloucester Laboratory. Frozen fish fillets, packed in an insulated shipping container, were shipped around the world by commercial air and ground transportation without added refrigeration. At the end of the trip, the product was still frozen. Because of their proven value, the use of insulated shipping containers, an idea introduced to the fishing industry in the 1970s, has grown considerably.

Needless to say, it is extremely important that the product be brought to the required temperature before containerizing it.

Time/Temperature Indicators

Since seafood spoils at a rate that depends on the temperature, and since the time for the products to spoil is fixed for any temperature or combination of temperatures to which they are exposed, it would be most useful to know how long a product has been exposed to any one temperature. With this information one can predict the remaining shelf life of the product. Unfortunately, this information is not available under ordinary circumstances, especially during shipment. It is to provide a remedy for this common shortcoming that time/temperature indicators have been developed (figure 9.11).

Sensitive chemical indicators change color when the product is exposed to a specific serial combination of time and temperature. The change in color is irreversible and gives an indirect indication that a portion or all of the shelf life of the product has been used up. The indicators are tailored by the manufacturer to integrate whatever combination of time and temperature information is required. They provide relatively accurate information, regardless of the time/temperature combinations to which the product may be exposed.

The cost of sending a chemical monitor with each container holding about nine to eighteen kilograms (twenty to forty pounds) is a pittance.

Figure 9.11. A Time/Temperature Indicator. (Courtesy Allied Corp.)

It is recommended that time/temperature indicators be considered for use in a quality assurance program as a means of checking expected performance. Where no quality assurance program exists, they are necessary to provide essential information that would not otherwise be available.

ADDITIONAL COMMENTS

Seafood processors should consider being part of a quality assurance team; of all of those involved in the handling of seafood, it is the processor who is identified either on the package or in official papers related to commercial transactions. Thus, the processor, unlike the fisherman, the retailer, and other handlers, has at stake his reputation and possibly even his financial security.

REFERENCES

Althouse, A. D., G. H. Turnquist, and A. F. Bracciano. 1979. *Modern Refrigeration and Air Conditioning*. South Holland, IL: The Goodheart-Wilcox Co. Inc.

Carver, J. H. 1975. Vacuum cooling and thawing fishery products. *Mar. Fish. Rev.* 37(7):15–21.

Carver, J. H., T. J. Connors, and J. W. Slavin. 1969. Irradiation of fish at sea. In *Freezing and Irradiation of Fish*. R. Kreuzer (Editor). Fishing News (Books) Ltd. London. pp. 509–513.

Gorga, C. 1983. A proposal for a feasibility study for the R & D of a new concept to maintain quality control at minimal cost through a cold chain system within processing plants. A proposal submitted to the NE Regional Office of the National Marine Fisheries Service, April 1.

Gould, W. A. 1977. *Food Quality Assurance*. Westport, CT: AVI Publishing Co.

Kramer, A., and B. A. Twigg. 1970. *Quality Control for the Food Industry, Vol. 1*, 3rd ed. Westport, CT: AVI Publishing Co.

Kramer, A., and B. A. Twigg. 1973. *Quality Control for the Food Industry, Vol. 2*, 3rd ed. Westport, CT: AVI Publishing Co.

Kreuzer, R. (Editor). 1969. *Freezing and Irradiation of Fish*. London, England: Fishing News (Books) Ltd.

Learson, R. J. 1984. Microwave thawing of fish. Personal communication. Gloucester Laboratory, Emerson Ave., Gloucester, MA.

Nickerson, J. T. R., J. J. Licciardello, and L. J. Ronsivalli. 1983. Radurization and radicidation: Fish and shellfish. In *Preservation of Food by Ionizing Radiation, Vol. III*. E. S. Josephson and M. S. Peterson (Editors). Boca Raton, FL: C. R. C. Press.

Nickerson, J. T. R., and L. J. Ronsivalli. 1979. High quality frozen seafoods: The need and the potential in the United States. *Mar. Fish. Rev.* 41(4):1–7.

Racicot, L. D., R. C. Lundstrom, K. A. Wilhelm, E. M. Ravesi, and J. J. Licciardello. 1984. Effect of oxidizing and reducing agents on trimethylamine n-oxide-demethylase activity in red hake muscle. *J. Agric. and Food Chem.* 32(3):459–464.

Ronsivalli, L. J., and D. W. Baker. 1981. Low temperature preservation of seafoods: A review. *Mar. Fish. Rev.* 43(4):1–15.

Ronsivalli, L. J., and R. J. Learson. 1973. Dehydration of fishery products. In *Food Dehydration, Vol. 2*. W. B. Van Arsdel, M. J. Copley, and A. I. Morgan (Editors). Westport, CT: AVI Publishing Co.

Slavin, J. W., J. H. Carver, T. J. Connors, and L. J. Ronsivalli. 1966. *Shipboard Irradiation Studies*. Annual report of the Gloucester Laboratory to the U.S. Atomic Energy Commission No. TID-23398 for period May 15, 1965 to May 14, 1966.

CHAPTER

10

The Roles of the Retailer and the Consumer

It is important that high product quality be maintained through distribution, retail sale, and finally preparation for consumption. The following chapter describes actions that retailers and consumers may take to assure high-quality seafood.

THE ROLE OF THE RETAILER

The retailer is the last commercial member in the primary chain of distribution of seafood. As such, his role is quite pivotal. A retailer, committing himself to a genuine attempt to be part of a seafood distribution team that assures the quality of its products to the consumer, must make certain that each and every seafood unit offered for sale in his store is of high quality.

He can give this assurance by following a few, but essential, common-sense practices. He should buy only those products that have enough reserve quality to last through the time that they will remain in the store and for at least one more day to be spent at the home of the customer. Since many processors put a "sell by" date on their product labels, this part of the retailer's obligation should not be too difficult to fulfill. Once the product is received, it should be transferred to the refrigerated display case as soon as possible; that portion which cannot be displayed immediately should be transferred to a refrigerated holding room. Fresh (unfrozen) products should be stored at about −1° C (30° F) and frozen products at about -29° C (-20° F). The temperatures of the display cases should be the same as those of the holding rooms, but in no case should they be higher than 0° C (32° F) for fresh fish and −17.8° C (0° F) for frozen fish. The first product in should be the first product out.

If the retailer adheres strictly to these simple rules, he will have done all he is expected to do in order to assure the quality of seafood to consumers. There are, however, two major impediments of the performance of his duties. One concerns the policy in relation to discards; the other, the design of the display case.

Discards Policy

Supermarket retailers seem to lack a policy concerning discards, therefore the basic rules are suggested here. When fresh fish have not been sold by the expiration date minus one (one day before the sell date), the retailer should do one of two things: (1) put them on sale in order to move them out of the store that very day; (2) freeze them and sell them as frozen products.

When fresh fish have not been sold by the expiration date, the retailer should do only one thing: discard them. He should not freeze them; he should not mark them down in price; he should not give them away. A consumer should not be allowed to obtain seafood of poor or mediocre quality under any circumstances!

Needless to say, a retailer can be successful in applying such a strict policy concerning discards only if he is sure of the expiration date. And in the retailing of seafood, this surety is generally obtained only by participating in a quality assurance program.

Seafood Display Cases

Major problems associated with retailers' inability to assure the quality of seafood products stem directly from the cases currently used to display seafood. One problem is inadequate temperature control, and the other is ineffective product rotation. Both are nearly unavoidable, given the designs of conventional display cases.

Conventional Display Cases. Some frozen foods are displayed in open vertical cases in which cold air is allowed to cascade from the top of the unit (figure 10.1). This method wastes considerable energy, since much of the cold air leaves the unit and needlessly lowers the temperature of the aisles; yet, although customers are exposed to uncomfortably cold air, much of the product is not maintained at the desired cold

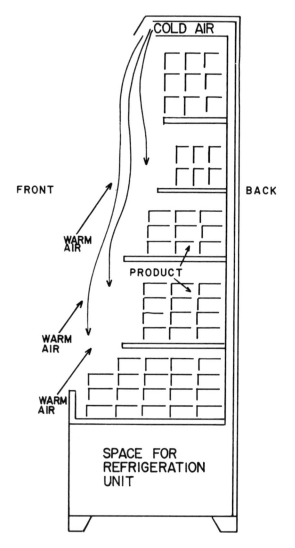

Figure 10.1. Open Vertical Display Case for Frozen Foods. (From Nickerson and Ronsivalli 1979.)

temperature. Some improvement is offered by those vertical units that are enclosed by glass doors (figure 10.2) or by a transparent plastic sheet that is slit vertically every few inches. These units, through which customers can see the products, retain the cold air and prevent the entry of warm air. At the same time, they allow easy access to the food.

Another type, the freezer bin with an open top, is of early design but is still used to some extent (figure 10.3). Since cold air is dense, very

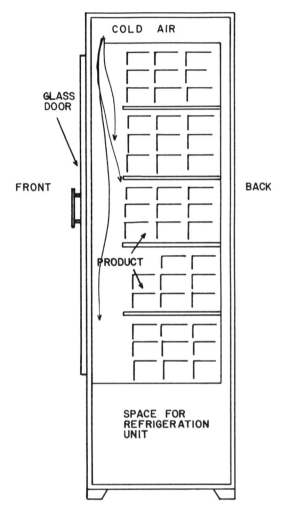

Figure 10.2. Closed Vertical Display Case for Frozen Foods. (From Nickerson and Ronsivalli 1979.)

little escapes this type of a unit: The only escape route for the air is up, and the denser colder air tends to go downward. But there are at least three problems associated with this type of display case:

1. They may be filled above the recommended load line, and then the uppermost products will undergo temperature rises and may even become thawed.
2. Considerable amounts of frost tend to accumulate on the product due to the unrestricted condensation of atmospheric moisture carried by

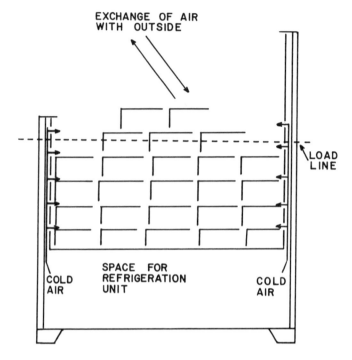

Figure 10.3. Open-Top Bin-Type Display Case for Frozen Foods. (From Nickerson and Ronsivalli 1979.)

the warmer air above the box, and this condensation may have an adverse effect on the salability of the product.

3. Condensation of moisture from the air adds heat to the system, and just as the system loses heat when it loses water by evaporation, so it gains heat when it gains water by condensation.

Lack of proper rotation—the second problem mentioned above— while more severe with the open bin case, is a general problem encountered in the selling of frozen seafood at the retail level with existing display cases. Regardless of the type of display case used, the FIFO principle (the first product in the display case should be the first one to be sold) may not always be observed because, no matter how diligently the retailer may set the sequence of the products in order to move the older products first, indiscriminate rearranging by rummaging shoppers usually undoes the retailer's best efforts.

The Proposed Seafood Dispenser. A dispenser type of display case appears to be the most promising, particularly since such a case represents one simultaneous solution to the two problems. The idea originated at the Gloucester Laboratory, where a prototype has undergone a series of preliminary tests. Even though the results of the tests have demonstrated the effectiveness of the method, it is evident that the current prototype design requires modification not of the concept but of the mechanics of its operation.

In its current design, it resembles other vending machines, except that it dispenses prepackaged frozen fish fillets without the need for coins (see figure 10.4 for diagram of product flow). It is envisioned that such machines would be placed in supermarket areas now occupied by the freezer display cases currently used for frozen seafood. The machine can be loaded with prepackaged products that have the price on the package label. Packages obtained from these machines would then be paid for at the cash registers. By releasing products from continuous tracks, the vending machine guarantees the first in first out principle; and the temperature control is facilitated because the system is closed and insulated from ambient heat. In addition, the release of product is through a small opening, which remains open for only a short period of time, thus reducing to a minimum the loss of cold air and the gain of heated air. Consequently, this type of unit reduces the energy requirements to display a given line of frozen seafood products and to maintain their temperature.

Considering its theoretical advantages, it is conceivable that this type of unit could eventually be used to dispense fresh seafood as well as other fresh and frozen perishable foods. It could also be used independently of the large retail markets and, instead, be installed either in small convenience-type stores or wherever vending machines sell frozen dairy products and other perishables. In the latter case, the machine would be coin-operated, and supplying and servicing the machine could be done by delivery truck drivers, as is currently done for products sold from vending machines.

Other Functions

In addition to his role of assuring the quality of seafood, the retailer is in a position to help consumers do their part so that the seafood they purchase will be of high quality at the moment of consumption. Retailers can provide literature on how to handle, store, and prepare seafood. They can also advise consumers about exercising care with perishables—not only seafood but also other meat and dairy products.

There is much work to be done in the area of consumer information and assistance. Some customers, for instance, are not aware that high-quality seafood does not smell fishy; it does not smell up the house during preparation. Most buyers seem to attribute the perceived differ-

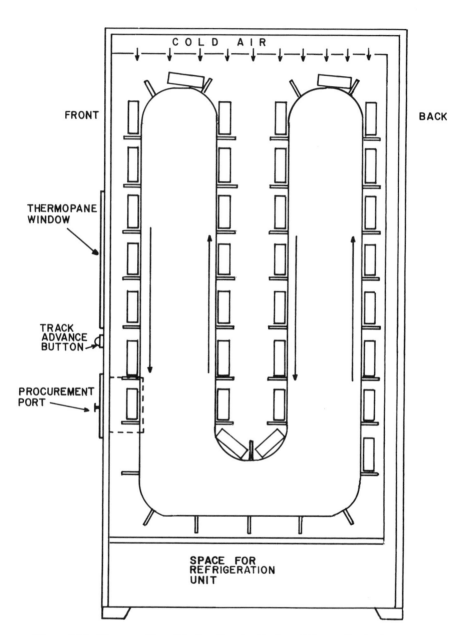

Figure 10.4. Dispenser-Type Display Case for Fresh and Frozen Foods. (From Nickerson and Ronsivalli 1979.)

ence between the quality of seafood purchased at supermarkets and that served in restaurants to their own deficient cooking and preparation techniques. They do not know that it is only fish of low quality that smells fishy, and, thus, they still have tolerance for fish of mediocre to poor quality. These attitudes, of course, do not foster any increase in per capita consumption rates, as they apply to domestically prepared seafood, and if the retailer fails to educate the consumer, he also fails to capture the economic potential that can be obtained from the sale of high-quality seafood.

Expected Benefits

By fully respecting the commonsense practices described above, adopting a strict policy regarding discards, and introducing the dispenser type of display case mentioned earlier, the retailer will be able to assure the quality of seafood to the consumer and, thus, will solve the major problems that exist today in the distribution of seafood through supermarkets.

One can describe these problems in a number of different ways, but they can be reduced to two interlocking ones: When the quality is poor, consumers remain uneducated about the possible delights that they might experience by eating seafood and consequently the per capita consumption of seafood remains low.

Three major effects can be expected when the national per capita consumption rates for seafood are raised: Through the establishment of a nationwide policy that assures the quality of seafoods, the country as a whole benefits from the exploitation of the maximum economic potential from its marine food resources, the business community benefits from economic growth as well as improved image, and the consumer benefits from better health. Not to be minimized is the wider selection of protein foods that can be offered to the consumer. It has, in fact, been amply demonstrated that the per capita consumption of seafood does increase as a consequence of this type of program. Finally, it should be noted that while a number of marketing strategies can be applied to stimulate a demand for seafood, their effect is a short-term one. The assurance of the product's quality is the only strategy that can be relied upon to provide a continuing increased demand for seafood.

THE ROLE OF THE CONSUMER

You may recall the emphasis placed throughout this book on the need for seafood to have enough reserve quality to last until the time it is consumed. Obviously, the consumer also has a role in the assurance of the quality of seafood. The consumer's responsibility is discharged with the performance of a number of commonsense activities that start with the shopping trip itself. If the trip includes other errands, food shopping should be the last. At the food store, the nonperishables should be selected first and the perishables last. Seafood may be either fresh or frozen at the time of purchase, but consumers should get acquainted with the criteria, pointed out in chapter 4, by which they can tell something about the quality of the seafood products. In addition to paying attention to those signs of high quality, the consumer should look for an expiration date on the package purchased.

Other important elements to be taken into consideration by the consumer are whether or not the product was produced under inspection and/or officially graded for quality. With inspected products, labels of inspection and quality grading will always appear on the package and are often imprinted on the producer's label. If neither the inspection nor the grade sticker is on the package, it means that the product was neither inspected nor graded by officially authorized personnel. It should be noted that it is entirely possible for a consumer to purchase the highest-quality seafood that may have been neither inspected nor graded. Many conscientious seafood retailers buy their products from equally conscientious seafood processors who do not process under the Federal Inspection Program. Consumers can buy from such vendors with confidence. However, unless they buy from a reliable vendor, it is strongly recommended that they purchase only those seafood products that carry the official inspection and grading stickers of the U.S. government. The reasons for this recommendation go beyond those related to organoleptic quality to include sanitary handling and other public health aspects.

The selection of a seafood item for purchase should be based more on the quality of the product than on the species. The consumer should not be reluctant to try new species or new products. From time to time, attempts by the seafood industry to popularize so-called underdeveloped or underutilized species have resulted in the availability of inexpensive products that offer eating qualities as good as, if not better than, the more popular species: At one time or another many of the now popular species were either underutilized or not used at all. As pointed out in chapter 2, these include the American lobster, halibut, haddock, Atlantic pollock, Atlantic red crab, northern shrimp, mahogany quahog, Atlantic whiting, and others.

A few of these species, such as the lobster, have generated such a demand that they are now among the highest priced of all species. But others, although achieving a reasonable level of demand, can still be found at significantly lower prices than the more popular species, whose organoleptic qualities are often not that much more desirable. For example, the fillets of Atlantic pollock may at times sell for approximately half the price of haddock fillets. Yet, pollock fillets can be prepared in any recipe that calls for haddock or cod fillets; the result is a comparable or nearly comparable product. Indeed, for instance, Atlantic pollock may be a more desirable species than many others for making fish chowder. It is safe to say that the major difference in eating seafood is not the species but the quality of the seafood, and, of course, the quality of preparation.

Because neither the shopping cart nor the automobile (or other forms of transportation) is refrigerated, no time should be lost in getting seafood purchases home and into the refrigerator or freezer. During the cold months of the year, foods may be better placed in the trunk space of the car; but during the hot months, they should not be put there unless they are protected by an insulated container with ice or other refrigerant.

When foods are brought into the home, the most perishable should be put away first and the least perishable last. Frozen foods should be held at as low a temperature as possible, but no higher than $-17.8°$ C ($0°$ F). The packages in which frozen foods are purchased are usually satisfactory for continued use, so there may not be a need to repackage them when the intent is to keep them frozen. If fresh seafood is to be frozen for later use, however, it may be necessary to repackage it, even if the fresh fish has been packaged when purchased: There may be an air space between the package and the fish, and whereas this is not a problem in fresh fish, it is in frozen fish. The product will undergo dehydration at the sites where the air spaces exist. The package to be used should not permit the transmission of water vapor. If the product is a fatty species, such as mackerel or salmon, or even if it is a less fatty one, such as swordfish and smelt, the packaging material should also be impermeable to oxygen in order to prevent rancidity.

Unless fresh fish fillets are to be prepared within minutes, they should be put into a refrigerator or freezer immediately. If they are to be held for up to a few hours, they may be refrigerated, but if held in a refrigerator overnight, they should be placed on a bed of ice in a leak-proof container because most domestic refrigerators are not held at $0°$ C ($32°$ F). Fish to be stored for more than twenty-four hours should be frozen.

The most desirable method for thawing frozen fish is with a microwave oven; the next most desirable is by placing it in the refrigerator (several hours to overnight).

The purchase of high-quality seafood and the protection of its quality until it is prepared for serving does not guarantee that it will be of the highest eating quality, unless it is properly prepared. This is the final responsibility of the consumer in assuring the eating quality of seafood.

REFERENCES

Anon. 1976. *The Taste of Gloucester: A Fisherman's Wife Cooks.* Gloucester, MA: Gloucester Cookbook Committee.

Nickerson, J. T. R., and L. J. Ronsivalli. 1979. High quality frozen seafoods: The need and the potential in the United States. *Mar. Fish. Rev.* 41(4):1–7.

Ronsivalli, L. J., and D. W. Baker. 1981. Low temperature preservation of seafoods: A review. *Mar. Fish. Rev.* 43(4):1–15.

ADMINISTRATION AND ECONOMICS OF QUALITY ASSURANCE

CHAPTER
11

Planning and Coordination

Although our definition of assured quality has been stated in various parts of the book, it cannot be repeated often enough that the quality of seafood cannot be considered to be assured simply because the product is of the highest quality when landed, when processed, or when sold. The quality of seafood can be considered to have been assured only when the product meets the highest standard for quality at the time it is consumed. It is then, and only then, that the ultimate judgment is made—a judgment that reflects on the reputation of seafood in relation to other foods and on the reputation and performance of the seafood industry as a whole.

In earlier chapters we discussed the characteristics of seafood, the deterioration and measurement of its quality, and the procedures used to assure the preservation of quality. Here, we shall describe the planning and administration we believe are necessary to assure the quality of all seafood sold. Keep in mind that it is necessary to differentiate between quality control and quality assurance. Quality control covers a range of activities that minimizes the loss of quality, no matter what the level of initial quality is. Quality assurance is an achievement that can be attained only when the initial quality of a seafood is high and remains so up to the moment of consumption. In order to assure quality, there must be an integrated set of activities and conditions that theoretically eliminates any probability that quality assurance will not be attained. More specifically still, to assure quality there must be planning, coordination, and, obviously, implementation of activities planned.

STRATEGIC PLANNING

Planning, coordination, and implementation of activities planned are indispensable elements for the achievement of nearly any goal. These elements were recognized as essential early in the conceptual development of the program of action to assure the quality of seafood to the consumer, a program that was designed by the Gloucester Laboratory. Seafood of high quality had traditionally been sold, especially along the coasts, by restaurants and seafood specialty stores. Two major strategic decisions were then made: to distribute seafood through supermarkets along the Atlantic coast and in the hinterland, and still assure the consumer of the purchase of a high-quality item. The distribution of seafood of assured quality was expanded by the participating firms, independent of the Gloucester Laboratory experiments, to specialty stores and restaurants throughout the country.

With the advent of supermarkets as the major merchandising channel of foodstuff in the United States and gradually throughout the rest of the world, to be out of supermarkets is to be out of the market for the middle class, namely the bulk of sales. But it is not easy to sell products through supermarkets. They do not change to accommodate products; it is products that change to accommodate them. This generalization was especially valid until the early 1970s in relation to seafood. Supermarkets tended to shy away from these products; because of the relatively small amounts sold, seafood was generally handled by the meat departments (rather than in-house specialized seafood departments, as is often done today); and it was treated with less than enthusiasm, especially because tools and equipment had to be cleaned every time fish interrupted the normal cutting and packaging lines of meat.

With these facts in mind, Gloucester Laboratory scientists suggested a number of changes that would bring high-quality seafood into the supermarkets, even in the hinterland. Most suggestions were adopted by the industry. The essential elements in the strategic planning to meet the requirements of a much longer chain of distribution, one with its own peculiar characteristics, were the development of a schedule of activities, a timetable to be respected by all members of the industry; the introduction of prepackaging at the processor rather than the retail level; inspection; and grading. Thereafter, a set of other minor changes were also incorporated in the program.

The Schedule of Activities

The very core of the schedule of activities (table 11.1) is that since the assurance of the quality of seafood depends on the performance of a

Table 11.1 The Shelf Life of Fish Fillets at Selected Temperatures

Temperature			
°C	°F	Shelf Life	High-Quality Shelf Life
26.7	80	1.0 day	.5 day
15.6	60	2.5 days	1.5 days
5.6	42	6.0 days	3.5 days
0	32	2.0 weeks	8–9 days
−1.7	29	3–4 weeks	15.0 days
−12.2	10	c 2.0 months	c 5.0 weeks
−17.8	0	c 1.0 year	c 7.0 months
−23.3	−10	c 2.0 years	c 14.0 months
−28.9	−20	>2.0 years	> 14.0 months
−40.0	−40	Several years	—

Note: c = about; > = more than.

number of elements in the chain of distribution, the degree to which quality can be assured is only the level that can be achieved by the least effective link in the chain. Therefore, regardless of the efforts that might be expended by any or most of the seafood-handling elements, quality will not be assured unless *all* of the elements are effectively coordinated to fulfill their prescribed functions.

Prepackaging

The next most important element in the strategic planning to sell fresh seafood of high quality through supermarkets is prepackaging. Most supermarkets have neither the personnel nor the expertise to care for the special requirements of unpackaged seafood. Unless they adopt the same strategy as seafood specialty stores, most supermarkets prefer to receive seafood in prepackaged trays. It is therefore suggested that packaging of fresh fish occur at the processor level. Expanded-plastic (a form in which density is reduced by increasing size primarily with air space, like a sponge) trays containing absorbent pads can be used; these trays are machine-overwrapped with self-adhering plastic film.

Inspection

As already emphasized a number of times throughout this book, unless one starts with seafood of the highest possible quality, the consumer is not likely to get a high-quality product. But how can we be sure that the quality level of the product remains high throughout the chain? Inspection has become one of the key elements in the over-all plan of action. Some of the functions that should be performed by inspectors have been described in earlier sections of this book; others will be described below in the section titled "The Mechanics of Coordination."

Grading

Once the product has been determined to be of the highest possible quality, it should also be certified as such. Hence the necessity for a grading system. For the quality assurance program, it was decided that only a quality level comparable to that of U.S. Grade A was acceptable.

Other Elements

Four other elements might be mentioned as completing the strategic plan of action that was designed and implemented in order to assure the sale of high-quality seafood to the consumer in supermarkets.

Pull Dates. The use of a clear marking stating the pull dates (set at a day before the product is expected to lose its high-quality characteristics) assures not only the collaboration of store personnel but especially the collaboration of the consumer in the implementation of the program.

A Special Label. This label specifies that the product has been inspected, and that it is of high quality (U.S. Grade A) at the moment of the inspection. The label also identifies the name of the processor who packaged the product and is participating in the quality assurance program.

A Trademark. It is expedient, if not necessary, to identify the products associated with any quality assurance program. In the case of a private company, this consideration is unnecessary because the manufacturer's name suffices. In quality assurance programs operated by states and cities, respective official seals might be used. Where a trademark does not exist, some means of identifying the products should be devised. The trademark serves several purposes. It identifies the product for consumers, thus guiding them in the selection of quality-assured products. It protects the integrity of the program by shielding it from any association with products that do not bear the trademark and that may be of poor quality. The trademark proclaims a cause-and-effect relationship between the dedicated efforts of the program's participants and the line of products—superior products—thereby produced. It also identifies the product for inspectors and retailers, so that they may perform prescribed functions on that product (e.g., inspection and rotation).

Education Campaign. The campaign of education of the consumer about the essential characteristics of high-quality seafood and about seafood preparation was barely hinted at in the beginning. This is one function of the overall program that has been intensified ever since by a number of industry members in various parts of the country, notably by the Fishermen's Wives Associations. It offers hopes of considerable payoff.

COORDINATION

To resolve problems that can be expected to arise from time to time, and especially to assist all participants in the quality assurance program in the fulfillment of their respective responsibilities, coordination of the program is indispensable. But who is to coordinate the program and how is coordination brought about?

How Coordination Gets Started

The establishment of a coordinated program to assure the quality of seafood begins with an earnest commitment to the concept of quality assurance. Such a commitment usually follows a comprehensive evaluation of the factors that influence the economic status of seafood production. It may be made by any organization within the seafood industry, fishery associations, and fishery cooperatives found throughout the coastal areas in the United States. It may be made at the government level, as was done in the state of Maine and the city of New Bedford, Massachusetts, at the management level by individual companies, or at the individual level by an entrepreneur who sees the whole program as a worthy investment.

Some Outside Assistance

Once the commitment to establish a quality assurance program has been made, we suggest that the organizers avail themselves of the considerable amount of technical information and data that are available from utilization-oriented facilities of the National Marine Fisheries Service, and especially the Inspection Service and the Gloucester Laboratory. Much technical assistance is also available from other agencies of the federal government. Provided that the information generated by a quality assurance program is made available to others, there is also the possibility of financial assistance from the government because of the potential benefits that may accrue to the nation. Because of the government's experience in the quality assurance of seafood, another possibility is to organize a cooperative pilot project with any of the NMFS utilization laboratories, with or without financial assistance. Cooperative projects with the federal government normally run for a period of one year, with occasional projects gaining an extension of one or more years, after which time the financial part of the government's assistance theoretically is no longer required. The technical assistance would probably continue to be available as needed.

In cases where the specifics of a quality assurance program, as well as the information generated from it, are to be kept confidential, neither federal financial assistance nor any cooperative effort with a federal facility can be contemplated. The technical data and other relevant information already developed by the government, however, are available at no cost and without restrictions.

In cases where the specifics of the operation are to be kept confidential, one can employ the "as needed" services of a private consultant. A large listing of food consultants is included in the annual *Directory of the Institute of Food Technologists*, which is the national society of food scientists and food technologists.

The Coordinator

Quality assurance coordination can be undertaken by an independent agency, by aggressive processors, by companies that are partially integrated vertically, or by state and local governments.

However selected, ideally an effective coordinator should possess a number of skills. In reality this person will have only some of them and will rely on associates who have additional skills. While the coordinator may not be required to have a degree in business economics, an effective one should be aware of the economic aspects of the duties and the economic goal of the quality assurance effort. And it is the benefits-to-cost ratio of any effort that the program undertakes or considers to undertake that should guide the coordinator so that benefits resulting from all efforts might be the highest for the least cost to the program. The coordinator must possess good management and supervisory skills, especially tact and resourcefulness, because this role involves the guidance and supervision of the activities of a substantial number of people, each having different motivations and operating in diverse work environments. The coordinator should also have some training in food science, because the program deals with a food commodity, and should have a working knowledge of sanitary (and the consequences of unsanitary) handling procedures, microbiology, food laws, packaging, chemistry, and toxicology.

Coordination may be a relatively simple task when the quality assurance program involves only the elements of a vertically integrated company. However, the majority of seafood entrepreneurs, particularly those dealing in fresh seafood, are too small to consider expanding vertically and too independent to become a part of a larger company. A quality assurance program for these smaller companies might be appropriately coordinated by a private organization established specifically to perform this function. Financial support for the organization might be derived from fees to be charged to the participating processors, perhaps on the basis of production volume. The fee is estimated to be a few cents per pound and can be included, as can all other direct and indirect costs of production, in the sales price to the consumer. The economic aspects of the overall program will be covered in detail in chapters 12 and 13, but at this point there should be no cause for concern that raising the retail price of seafood a few pennies per pound might affect the demand. The added cost is insignificant since retail prices of seafood—particularly fresh seafood—fluctuate from week to week by amounts that normally far exceed that cost. As against such a negligible increase in price, the benefits of the program to the consumer are consistently and conspicuously high.

The Mechanics of Coordination

It has been emphasized that a quality assurance program must be coordinated if it is to be successful. We have already addressed two general issues related to coordination: who should perform the coordinating functions and how does a coordinated program of quality assurance start. This section will cover the process by which a successful program operates.

Rules and Roles. The coordinator is a rule giver. Each participant in a quality assurance program must, in fact, play a specific role, in accordance with specific rules (e.g., time allocation and sanitary practices) that clearly specify what the responsibilities to be fulfilled are. Initially, the coordinator will set down preliminary rules, which, although based on facts and experience, should be considered temporary. With practice, reasonable modifications may be introduced in order to improve the conditions under which one or more of the participants must operate. However, no modifications should be made or even considered if they might jeopardize the assurance of the quality of the product. In setting up the rules under which participants must operate, the coordinator can rely on an advisory committee representing the participating elements of the industry (e.g., a retailer, a processor, and a fisherman).

On an annual or semiannual basis, the coordinator should obtain consumers' points of view concerning specific products and the program as a whole, whether through ad hoc meetings, formal surveys, or informal interviews at the retail site. Using these and other strategies, the coordinator should be able to oversee the program in ways that benefit all concerned.

Supervision of the Program. The coordinator acts as a supervisor, formulating the rules and taking the necessary actions to assure that they are indeed observed by all concerned. It is not necessary to emphasize that the coordinator is not a czar or a dictator, but merely one of the program's participants. As any effective supervisor, the coordinator must follow the same rules that result in good and effective supervision and management in any type of business operation. Although the rules of good management are outside the scope of this book, an old but enduring publication on the subject is cited in the references (King 1944).

It is up to the coordinator to make sure that all elements of the handling chain fulfill their prescribed responsibilities through a set of activities, including communication, research and monitoring of the quality of the products, the status of participating facilities, the performance of participants, and other related duties. Because of their intrinsic importance, some of these are examined in some detail.

Communication. The coordinator makes sure that all participants in the quality assurance program are made aware of the rules of operation and that no ambiguity of interpretation is left in the mind of any participant. The coordinator also assembles and distributes to all participants available information that is relevant to the operation of the program. The coordinator might issue a short monthly report as a vehicle for transmitting educational and other pertinent information to all participants in the program.

Research. A major area of research for the coordinator lies in the continuing investigation of the time allocations to the participants and the times during which the products remain at high quality. Those suggested in this book are provided in table 11.2. It must be stressed anew, however, that the relevant data supplied here represent only a starting point and should be modified as the program develops, especially respecting local conditions.

Beyond the schedule of activities for participants, there are numerous problems connected with the conservation of the quality of seafood. As a consequence, there is a considerable amount of research effort that is continually expended at government (federal and state), academic, and industrial laboratories. Results of this work may have a direct or indirect relationship with the quality assurance of seafood. As an example of

Table 11.2 Time Allowed Each Element in the Distribution of Fish Fillets

Distribution Elements	Maximum Time Product Can be Held	
	At 0°C (32°F)	At −17.8°C (0°F)
Vessel	2 days	7 days
Processing plant	1 day	1–2 days
Warehouse	—	About 6 months
Retail outlet	5 days	About 3 months
Home	1 day	About 3 months
Total time	9 days	About 1 year

directly relevant developments, we would cite a hypothetical finding that lowering the storage temperature by 1° C (1.8° F) adds one day to the shelf life of unfrozen seafoods. (We hasten to emphasize that this is not being reported as a fact; it is merely a hypothetical example of directly relevant developments.) At the very least, this information is useful and may even be of vital importance to one or more of the participants.

An example of indirectly relevant developments is the hypothetical finding that a plastics manufacturer has developed a process for making puncture-proof plastic garbage bags. On the surface, this finding appears to have no relevance to the assurance of the quality of seafood. On the other hand, it may have relevance to such operations as the bagging of whole fish having sharp, sturdy spines, such as the Atlantic red fish, or of any of the many shellfish whose exoskeletal parts are hard and sharp.

It is essential that the coordinator stay abreast of such developments by maintaining a rapport with at least one research facility and visiting its library once a month to scan the current journals and periodicals. The NMFS technological laboratories and the NMFS Inspection Laboratory are typical examples of such facilities. If the coordinator is for any reason unable to give serious consideration to this recommendation, he should establish a relationship with the nearest university with a curriculum in fisheries or food science. Most of the state universities have one or the other, or both. Having established a rapport with either federal or academic food scientists, the coordinator may be in a position to attract the attention of these researchers to problems that are encountered in his program, particularly when their solutions have the potential for broad applications. In all circumstances, the involved and informed coordinator is bound to be more effective in his work, and accordingly will be more respected by the participants in the program.

Should the coordinator be unable to obtain the interest of federal or academic scientists, the possibility exists of gaining approval from his colleagues to allocate a small amount of money to investigate any specific problem encountered by the program. The money may be raised through a short-term increase in the fees to the participants. If, because of potential benefits, the decision is reached to investigate the problem, whenever appropriate, the coordinator will probably get the most results using the money to support either a graduate student at a university or a part-time employee at one of the government laboratories.

Monitoring. Perhaps the most important role of the coordinator is monitoring the program, including the performance of the participants, the facilities that are involved in the program, and especially the products, at every stage of their distribution.

The ease with which the coordinator is able to perform this role will, of course, vary, depending on numerous factors. As a general rule, after the optimum size of the operation has been reached, the difficulty of monitoring increases in direct proportion with any further increase in size. No matter how seriously management of such especially large concerns dedicates itself to a purpose, and no matter how strenuously it attempts to impart a similar dedication to the lower echelons, the evidence is that management does not always succeed. It seems to be the human tendency to stray from strict performance, even though one has made a seemingly firm commitment to it.

Monitoring will be least problematical in the processing plants, especially when, for any given volume of production, the plants are few in number, exceedingly efficient, and correspondingly of appropriate size—in addition to manifesting clear dedication to service. These are the firms that often grow from small to very large size. (It would not be fair to leave this topic of the relationship between size and efficiency with its impact upon the ease of monitoring without stating that there are exceptions to the rules: Small companies do not always provide desirable service and do not always grow to become large companies; similarly, companies that grow to immense proportions do not, in all cases, necessarily lose the essence of service that made them great.)

In large plants, the unit cost of monitoring is relatively small. Monitoring of the smaller plants and the retail outlets may require methods whereby costs can still be kept low without sacrificing the required performance. However, monitoring the fishing vessels and the actual fishing processes is apt to be the most difficult, to say the least, particularly because most fishing vessels are small and conditions that prevail at sea are severe and sometimes forbidding.

Because of these difficulties, it has been concluded that it is not economically feasible to keep most fishing vessels, small retail outlets, and small plants under continuous inspection. There is one characteristic of the smaller facilities, however, that tends to minimize the need for continuous inspection, and the smallest facilities may require only occasional monitoring. This characteristic is that many of the fishing vessels and most of the small processing plants are family-owned and family-operated, with little, if any, nonfamily help. In such cases, all members are generally motivated to enhance the position of the company in every respect. Once they are given the rules to follow, they will comply, knowing that noncompliance could jeopardize the economic position and image of their company. Thus, provided they meet detailed specifications and follow a set of operating instructions, even small facilities can be certified as participants in the program. Thereafter, periodic checks to see that the performance is as it should be would probably be adequate.

Without some form of monitoring of operations, there is bound to be a relaxation in performance, and nothing can be more damaging than a consumer's shaken confidence in the ability of the program to deliver assured quality products.

Using the Services of Inspectors. The total monitoring required in any program, regardless of size, cannot possibly be accomplished by the coordinator personally. While he has the ultimate responsibility for the monitoring of the program, the actual work will be done by others. Given this fact, the services of the United States Department of Commerce Inspection Service should be considered. We recommend this option for several reasons. The USDC Inspection Service is the only entity with authority to place official inspection and grade labels on seafood products; and these labels alone are recognized in legal proceedings. In addition, these labels alone are given official recognition in international trade and in purchases by federal, state, and local agencies, as well as other institutions. In fact, the official standards for seafood promulgated by the U.S. government and enforced by the FDA are prepared by this agency. Another important reason is that monitoring can be done in a number of acceptable ways. The government's seafood inspection service has determined the reliability of these different methods and can help the coordinator select those best suited to the facilities that must be monitored. Also, the government inspectors do more than just inspect the products. They are trained in the sanitary aspects of production and have current information about possible environmental contaminants. By the nature of their work and contacts, they have access to information on developments in quality control, quality testing methods, and any new problems associated with the wholesomeness of seafood. Finally, the federal government's inspection service is flexible.

It includes full-time or continuous inspection, part-time inspection, lot inspection, inspection and accreditation of facilities, and other services, so that a whole range of inspection needs may be accommodated.

It is for these reasons that we recommend that the coordinator maintain a rapport with the USDC Inspection Service, regardless of the extent to which the service is used by the program. It is entirely possible to establish an effective monitoring system without any government involvement, of course, but we believe that the probability of success would be reduced somewhat in such cases.

Special Assistance to Participants. The opportunities to improve seafood handling are many; all that is required is that the new as well as some older ideas be considered and tried by the industry. This is one area in which the coordinator should use a combination of technical background and knowledge of latest developments in order to solve problems or to improve the operation of the program. The participants should be made aware of potentially beneficial ideas, and should also be assisted in setting up and conducting pilot tests, in evaluating the results of the test, and in arriving at decisions on whether, and how, to proceed in the adoption of these ideas.

For example, whereas many fishermen make short fishing trips (one to two days), there are also a goodly number whose trips last a week or more. Thus, fishermen who have traditionally brought in iced fish as "old" as one week or more (in this case, the age of the fish is counted from the time of death, which usually occurs at or about the time it is caught), and processors who have distributed the ensuing products as fresh seafood are apt to be dismayed when told that only fish two days from catch will be allowed to enter the line of fresh products handled by the quality assurance program—especially when the retailer is a supermarket chain. The fishermen will become indignant at such an apparently unnecessary and drastic change in tradition, especially because they can see that the quality of the fish they bring in is high. Part of the indignation reflects the frustration at the prospect either of making a drastic change in their traditional fishing methods or of selling the older part of their catch to processors of frozen seafood at a lower price than could be gotten from selling it to processors of fresh seafood. Fishermen have to be educated to appreciate that, while the quality of one-week landed fish is high, it will *not* be high by the time it is consumed! That is the rationale that the coordinator must successfully convey. This example of a problem facing the coordinator is a real one, and it may come from fishermen and processors alike.

It is a demanding exercise for the coordinator. On the surface the problem appears to be insoluble, because the reaction reflects an adamant position and because the shelf life of high-quality seafood simply does not allow any compromise of the time allotted to fishermen and processors. The coordinator has to convince them of the facts. For example, any concession given the fishermen is taken from the processor—and vice versa, up and down the line. Above all, the coordinator must keep track of the products so that an allowance to land fish three or four days old, for instance, does not result in fish that are more than nine days old when consumed. The coordinator, however, has other alternatives for fishermen who cannot meet the two-day rule: Assuming that the catch is not sold to specialty stores and restaurants, the coordinator can suggest they freeze or superchill the older part of the catch, or that they salt it. But suggesting these alternatives is not enough. If the coordinator is to succeed, a sincere attitude must be demonstrated by working with them at eliminating or reversing the disparity in market prices between fresh and frozen fish, and by locating the markets for their frozen, superchilled, or salted fish.

Nor do challenges exist only for fishermen. One of the outstanding problems that exists in the processing plant is how to keep the product cold during the production process. The reasons for this problem, as well as a proposed solution employing a refrigerated conveyor system, the Friotube, were discussed in chapter 9. The coordinator can help the industry by working with one of the more aggressive and innovative processors to develop a prototype and to conduct pilot tests to determine the technical and economic feasibility of this proposed method, which can then be used to the benefit of all the participants in the program.

Another example of how the coordinator can assist the participants in the program involves what he can do to assist the retailers. As things stand now, in general, supermarket retailers do not have a satisfactory performance record in assuring the quality of fish sold to consumers. The coordinator can help the retailers improve their performance by convincing them to take on at least two critical responsibilities: (1) Never permit a customer to leave the store with a seafood product of less than U.S. Grade A quality; (2) all fresh fish that remains in the store until one day before the expiration date either should be put on sale for one day and then discarded if not sold by the end of that day, or should be frozen on the day prior to the expiration date, and then displayed as a frozen product until sold. Retailers should be cautioned against converting fresh fish to frozen fish on the date of expiration. By then, it is probably too late. The strategy of converting poor-quality fresh fish to frozen fish in order to avert a financial loss is probably the single outstanding practice that has led to the poor reputation and low retail price of frozen seafood. More specifically still, this set of recommendations implies that all products that fall below U.S. Grade

A quality must be discarded. They should not be frozen and sold as frozen products. They should not be marked down. They should not be given away. They should be discarded.

Two other outstanding problems associated with the retailing end of seafood are the inability to hold it at the desired temperature and to follow the first in, first out rule. The reasons for these shortcomings have been described in chapter 10 and a solution—using a new type of fail-proof display case—outlined. The coordinator could be instrumental in the solution of these problems either by working with the NMFS Gloucester Laboratory where a prototype model of such a display case exists and has undergone preliminary testing, or by borrowing the case and conducting pilot tests in cooperation with one of the retailers, or working with the manufacturer of the case on a second and improved prototype.

Other Duties of the Coordinator. Among the variety of tasks that the coordinator needs to perform are those of compiling and analyzing the data generated by the program. The data to be compiled include production rates, costs, sales, losses of product, etc. The analyses to be performed include the determination of costs and sales trends, the effects of program variables, the impact of problems, etc. The interpreted findings are used as a basis to restructure time allocations and to make other adjustments in the prescribed set of rules under which the program operates. All relevant costs are analyzed and benefits-to-cost ratios are determined, even if only estimates, as a basis for any recommendation to be offered.

The coordinator must maintain accurate records of the information generated by the program and its operation, such as listings of participants, facilities, and vessels, as well as records of important events related to the program. The coordinator should maintain a file of all relevant food laws, technical publications, tests for quality, sanitary methods, etc.

The coordinator should also keep all participants fully informed about the program's progress, maintaining close contact with all participants by making periodic visits to each facility in order to view the operations and to discuss problems and other relevant topics. He also carries out the public relations function of the program, the scope of which will depend on the program's evaluation.

How Many Coordinators. Needless to say, we speak of one coordinator only for ease of exposition. In reality, there is a need for as many coordinators as there are separate programs of production and distribution of high-quality seafood. The simplest program might involve a single processor, a more complex program, more than one—and perhaps an entire port. Potentially, there can be one coordinator for every major region of the country. Conceivable, but not to be recommended, is the procedure of establishing only one national coordinating office. Cost should, of course, be a major consideration in making these decisions, but so should the expected benefits.

IMPLEMENTATION OF A QUALITY ASSURANCE PROGRAM

Many aspects of the program of quality assurance of seafood described throughout this book were actually implemented in a federally coordinated program. The program started in the late months of 1975 with the delivery of about 200 pounds of fresh seafood per week from Empire Fish Co., Inc. in Gloucester, Massachusetts, first to stores of Great Atlantic and Pacific Tea Co., Inc. (the A&P supermarket chain) and then to stores of DeMoulas Super Markets, Inc. Both sets of stores were located in Massachusetts. By the end of the official pilot project in 1980, as we shall see in more detail in the next two chapters, the program became national in scope and sales grew to about 11 million pounds.

The motivation to implement this program on such a scale was largely based on the economic potential to be derived from it. From the viewpoint of the U.S. seafood industry, as we shall see in chapter 12, the quality assurance of seafood is expected to return satisfactory economic benefits due to expanded demand in new and old markets that are ready to pay a reasonably higher price for high-quality seafood. An ancillary, but highly important, benefit to be gained by the seafood industry is an enhanced business image. And these are not the only benefits. From a national viewpoint, as we shall see in chapter 13, direct economic benefits will accrue from a reduction in the difference between the value of seafood imports and exports, which currently accounts for a deficit of over $5.5 billion per year; indirect economic benefits will also accrue to the nation from the possibly improved health of the population, a result of the increase in the per capita consumption of seafood.

REFERENCES

Gorga, C., and L. J. Ronsivalli. 1983. Quality control and quality assurance: Getting the difference straight. *Infofish Marketing Digest* 4:32–34.

Guthrie, R. K. 1980. *Food Sanitation*, 2nd ed. Westport, CT: AVI Publishing Co.

Institute of Food Technologists, n.d. *1987 Directory*. Chicago, IL.

King, W. J. 1944. *The Unwritten Laws of Engineering*. New York, NY: American Society of Mechanical Engineers.

Ronsivalli, L. J. 1982. A recommended procedure for assuring the quality of fish fillets at point of consumption. *Mar. Fish. Rev.* 44(1):8–15.

Ronsivalli, L. J., and R. J. Learson. 1973. *Dehydration of fishery products*. In *Food Dehydration, Vol. 2*, 2nd ed. W. B. Van Arsdel, M. J. Copley, and A. I. Morgan (Editors). Westport, CT: AVI Publishing Co.

CHAPTER

12

The Economics of
Quality Assurance
of Seafood

There is one essential reason that most of the program of quality assurance of seafood described in this book was implemented in such a
smooth and swift fashion: The program was economically feasible and
it yielded a satisfactory financial return to all participants.

The returns to the processor and the retailer were calculated in some
detail and are reported below. Those of the fisherman and the consumer
were not calculated because, although considerable, they were indirect
and diffuse. Fishermen gained from an improved image as an industry, from an expanded market, and from the potential of eventually
sharing in the financial benefits that can be derived from the program.
Consumer gains included a larger choice of high-quality species available at convenient locations in the market (many more species have
gradually been introduced during the last few years and many more
stores, especially in the hinterland, have been carrying fresh seafood).

They have also gained in more specific ways. In the absence of quality
assurance, consumers who buy a product of poor quality are faced with
three choices: They can consume the product—and perhaps never buy
seafood again; they can discard the product themselves; or they can
return the product to the store, in which case they might be reimbursed
for the cost of the product but not for the time and cost of travel.
Consumers, therefore, incur either a direct or an indirect financial loss,
especially in the last two cases. Quite apart from all other benefits, it is

this financial loss—and the consequent likely aversion to ever buying seafood again—that is eliminated through a program of guaranteed high-quality sales.

The financial returns to the processor and the retailer were determined through detailed analyses as the program unfolded intermittently over a period of about five years. The procedure followed for the determination of those returns, as well as some of the specific findings of these analyses, are reported below. The study was conducted in four stages, three concerned with fresh fish fillets and the fourth, with frozen fish fillets. It must be kept in mind that before the Gloucester Laboratory of the National Marine Fisheries Service (NMFS) undertook this study, there was no available evidence regarding either the technological or the economic feasibility of such a program. It might, therefore, be wise to give some attention to the basic reasons for which the study was undertaken.

RATIONALE FOR THE STUDY

The study concerned with fresh fish was specifically conducted to determine (1) whether the technological conditions to achieve and to preserve the highest possible quality would be adopted by the industry; (2) to what degree the implementation of those conditions would raise production costs; and, finally (3) whether consumers would buy the product at the expected higher price, thus leaving a safe margin of profit for both the processor and the retailer (Gorga et al. 1979).

The need to extend the study to cover frozen fish fillets was dictated by an additional set of reasons (Gorga et al. 1982). First, although mainly imported, frozen fish fillets comprise the bulk of the fillet supply in the United States and suffer from most of the problems besetting the industry. The underlying questions to be tested, therefore, were: Is it possible to produce domestic frozen fish fillets that are of such high quality that they satisfy the requirements of the market, and are eventually able to replace the imported fillets? A second group of questions was dictated by a complex set of factors, which are illustrated in figure 12.1. This figure divides the entire market into two parts, one for fresh and the other for frozen fish fillets; but it does not depict their true relative size. The figure attempts to point out that there is a one-way permeable barrier between the two markets: Given a sufficient supply, fresh fish invade the frozen fish market, but frozen fish do not do the reverse. Therefore, the second underlying group of questions to be tested was: Can that barrier ever be destroyed so that frozen seafood will be able to enter the fresh fish market, and to perform the expected function of smoothing out the erratic swings that exist both in the

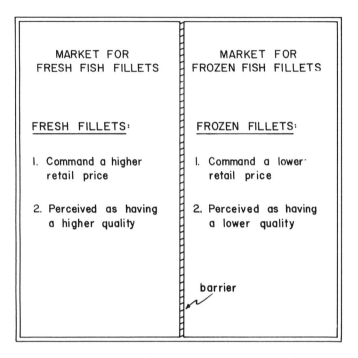

MARKET FOR FRESH FISH FILLETS		MARKET FOR FROZEN FISH FILLETS
FRESH FILLETS:		FROZEN FILLETS:

1. Command a higher
   retail price
		1. Command a lower·
		   retail price

2. Perceived as having
   a higher quality
		2. Perceived as having
		   a lower quality

	barrier	

Figure 12.1. One-Way Barrier between Fresh and Frozen Seafood Markets.

supply and the price of seafood? Ultimately, the question to be tested was whether the frozen product will ever be able to satisfy a difficult-to-measure, but evidently high, latent demand for seafood (Nickerson and Ronsivalli 1979).

Figure 12.1 also attempts to illustrate that since the barrier in the seafood market is the result of the interplay of two major factors—retail prices and perceived quality—these factors automatically become the keys to overcoming that barrier. Clearly, the element of relative retail price was not the answer. Otherwise, since frozen fillets were already sold at a lower price, they would have long ago invaded and destroyed the market for fresh fillets. What remained was the element of relative quality. Thus, the specific questions to be answered by the study were exactly the same as those concerning fresh seafood: (1) Was it technologically feasible to produce high-quality frozen fish fillets; (2) how much more would it cost to produce high-quality frozen fish fillets; and (3) would the consumer pay the expected higher price? Restaurants, including fast-food chains such as McDonald's, were already giving a resounding positive reply to these questions. But would the super-markets be able to become a comparable outlet to satisfy the growing demand for seafood?

AN OVERVIEW

The fresh fish pilot project run by the Gloucester Laboratory has clearly blossomed into a national program of considerable magnitude. Unfortunately, the same cannot be said for the frozen fish pilot project. For a variety of reasons, the recommendations associated with it have not taken root in the industry. Unless a major effort is made along the lines pointed out in this book, and reemphasized in figure 12.1, the situation will remain the same, because it is obviously difficult to alter the momentum under which it continues.

SPECIFICS OF THE STUDY

The study analyzed the results of four individual experiments: the first three for fresh fish, and the fourth for frozen fish. In all experiments, fish were processed in plants that were approved by the USDC Inspection Service. Individual fillets were placed in trays that were overwrapped by machine and heat-sealed in clear plastic film. Fillets that met the required quality criteria (see chapter 4) and passed the official set of handling, filleting, and packaging guidelines were classified as U.S. Grade A fillets. It was these fillets that were of special interest to the study. Labeled as U.S. Grade A and carrying the USDC Inspection sticker, the species name, the name of the processor, other appropriate logos, and, later, pull dates, individual packages were shipped to the retailer in master cartons whose temperature was controlled either by refrigerated vehicles or by dry ice (solid CO_2) during hot days. Each package was weighed and priced at the retail level.

The first three experiments involved a single processor in Gloucester, Massachusetts, and three different supermarket chains located in different areas of the commonwealth. Experiments I and II were carried out between October 1976 and February 1977; they helped determine the logistics and operating parameters of the program. Experiment III covered a twenty-one-week period between May and September 1977. In this experiment, there were two test stores and two control stores (those in which no assured quality fillets were sold). Data on deliveries and prices concerning the two control stores were provided by the retailer; other data were collected by laboratory personnel directly at the test stores or the plant level. Those on costs and cost estimates were obtained through specific studies.

Experiment IV covered a sixteen-week period between February and May 1981. This experiment involved one processor in Boston and one supermarket chain in the Albany/Schenectady area of New York State. This specific location was sought out for two major reasons. First, as pointed out above, even though retail prices tend to be lower for frozen fish, the cost of production is higher than that of fresh fish; therefore, an attempt was made to select stores away from the coast, where such an economic dysfunction is less likely to be intensely felt. This selection was also dictated by the presumption that sales of frozen fish are higher inland than along the coast. In this experiment, since the retailer soon introduced the product throughout the entire chain, control stores were automatically transformed into test stores. The resulting wider selection in relation to the location of the stores permitted a study of the correlation between sales of seafood and varying socioeconomic conditions. The five stores that participated in the experiment for the longest period were located along an axis that starts at the center of Albany and ends at the periphery of Schenectady (see figure 12.2 for a schematic representation of this geographic arrangement). This disposition is interesting because it is an approximate representation of various socioeconomic strata of store customers, from low to middle to upper income.

In this last experiment, unlike the previous ones on fresh fish, responsibility for the collection of relevant data—with the exception of spot checks for temperature—was assumed by private industry. Thus, data on production volumes and costs were collected by the processor and those for retail sales and consumer prices, by the supermarket chain.

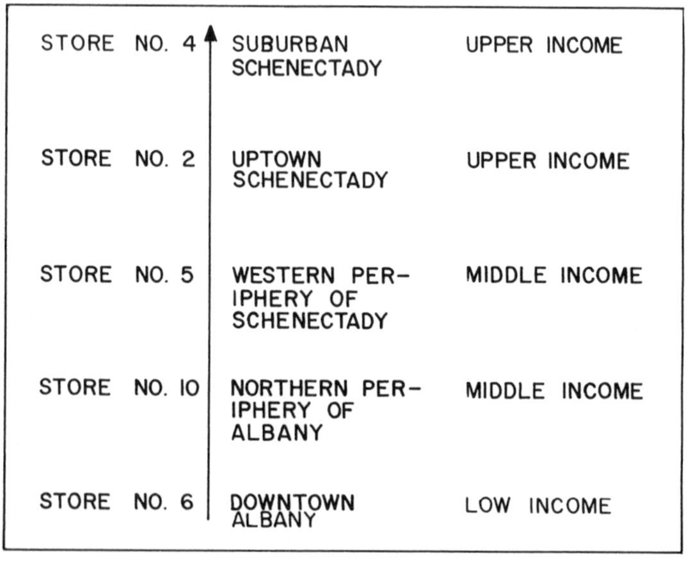

STORE NO. 4	SUBURBAN SCHENECTADY	UPPER INCOME
STORE NO. 2	UPTOWN SCHENECTADY	UPPER INCOME
STORE NO. 5	WESTERN PERIPHERY OF SCHENECTADY	MIDDLE INCOME
STORE NO. 10	NORTHERN PERIPHERY OF ALBANY	MIDDLE INCOME
STORE NO. 6	DOWNTOWN ALBANY	LOW INCOME

Figure 12.2. Schematic Representation of Geographic Location of Test Stores and Income Status of Customers.

For all experiments, data were collected by species: In addition to cod, haddock, flounder, and ocean perch, other species held under observation at times were pollock, whiting, and cusk. Wherever possible and necessary, data were also collected for substitute protein products, such as meat and poultry. All data were gathered at the processor level, in test stores, and in control stores for varying lengths of time and, frequently, for different time subdivisions.

Administration. In all experiments, processors decided on their selling prices, and store managers on their selling prices as well as on quantities and species ordered. Since some costs were assumed by federal support funds for the study, a few words are necessary to clarify the rationale for the use of those funds.

Federal Coverage of Special Costs. In order to reduce the economic risk to the industry and to obtain a true picture of the effect of quality assurance on the demand for quality seafood, federal support funds for the program covered costs of unsold product (returns), inspection, and, for the fresh fish experiment only, transportation. Costs of monitoring and analysis of the program were also covered by the support funds. Without these funds, it would not have been possible to undertake the study: Processors were reluctant to participate in an experiment whose added production costs might not be fully recoverable, and retailers had to be encouraged to overbuy in order to insure that the demand was fully satisfied rather than cut short by lack of supply or, even worse, by an erratically short supply.

PROCEDURE TO DETERMINE ECONOMIC FEASIBILITY

The reasoning used to determine the economic feasibility of the program is as follows. There are fixed costs in producing and selling fish of any quality. In producing and selling poor-quality seafood, indirect costs stem from a decline in sales and corresponding product loss due to spoilage. These combined factors increase the unit cost of the product because fewer units are sold, but the fixed costs of production remain the same. For high-quality seafood, on the other hand, indirect benefits accrue from increased sales and reduced product loss due to spoilage. These factors lower the unit cost of the product since more units are produced (and probably sold) per fixed costs of production.

No attempt was made to study indirect costs or benefits; the direct ones for the program as a whole, however, were determined as the *cost differential* between buying and selling fish of any quality (fish of not assured quality) and buying and selling fish of high quality (fish of assured quality). With the task concerning costs thus defined, it was possible to overcome one of the biggest problems that often plagues studies of this type: confidentiality of the data. Since we were not interested in obtaining information on the net profit margin of the economic operators involved in the study, but were interested only in the cost differential between buying and selling the two types of fish, we were able to obtain all the data that were needed. Specifically, since they vary from firm to firm and from time to time, we were not interested either in overhead costs such as administrative expenses, financial charges, or taxes paid by each firm. Ultimately, we were not even interested in the specific gross profit margin of each participant in the study; rather, we were interested in the gross profit margin of the ideal processor and retailer who had reached an efficient level of operations. We call these the prototype processor and retailer. For the processor, in particular, all costs were given and unit costs were estimated on the assumption that the production schedule had achieved a minimum efficiency level of 10,000 pounds per day. Three factors helped in determining this common denominator: inspection costs; the capacity of the wrapping machine; and transportation costs. No attempt was made to establish the upper limit of the efficiency level.

Determination of Added Costs To Assure Quality

In order to determine the cost differential for producing and selling fish of assured quality and fish whose quality was not assured, it was necessary to determine the cost involved in adopting specific technological recommendations that were expected to assure the high quality of the product at the moment of consumption. The specific cost differentials that were of interest concerned the respective costs of specialized filleting (more abundant and more careful trimming to assure the removal of bones, for instance) versus normal filleting; specialized packaging (preparation of individual packages with use of tray, absorbent pad, and plastic film over-wrap) versus bulk packing in tin cans and wax paper; the cost of the wrapping machine, which did not exist for other fillets; insulated master cartons versus wooden boxes; and cost of inspection, which did not exist for other fillets.

Most of these costs—such as the unit cost of packaging materials—were straightforward and could be easily calculated; others needed to be determined. The methodology and the specific formulas applied to derive those costs are described in appendix 5.

Gross Profit Margins

In order to obtain gross profit margins for the prototype processor and prototype retailer, three levels of prices were observed for each species of fish included in the study: ex-vessel, processor, and retail prices. Figure 12.3 reports these for only one species, fresh cod. (For this species, retail prices for the fifteenth and seventeenth week were lower than processor prices; for a variety of reasons, the retailer on those occasions sold the product at a loss, as a loss leader.) Using formulas that are described in appendix 5, operating costs were subtracted from selling prices for the processor and the retailer. As pointed out earlier, no overhead figures were calculated for this study, so only gross profit margins were obtained.

FINDINGS

The use of these methods and procedures yielded a number of specific findings. (Some of the broader ones are presented in the next chapter.) However, because of the differences in time, in the nature of the product, and in operators involved, we will report specific findings separately for each of the four experiments comprising the study.

Fresh Fish: Experiment I

Experiment I dealt primarily with the technological feasibility of producing U.S. Grade A fresh fish fillets, but also considered production and transportation costs.

It yielded very little hard data, but helped to determine the logistics and operating parameters of the program. Thus it was determined that—for a typical processor—it would take one day to prepare and to ship about 10,000 pounds of high-quality seafood, and that the technology for shipping seafood while keeping the temperature at nearly constant levels was adequate. It was also determined that if transportation would take more than one day, the wiser alternative was to ship the product by air rather than by land.

Fresh Fish: Experiment II

Experiment II dealt with the production, distribution and sale of relatively few pounds of fresh fish fillets concentrating on consumer reactions to quality and price.

Experiment II gave a strong indication that the suggested innovations were highly acceptable to the processor and the retailers, that consumers were ready to pay a higher price for fillets of assured quality than for fillets whose quality was not assured, and that sales tended to increase when the U.S. Grade A label appeared on the package and the

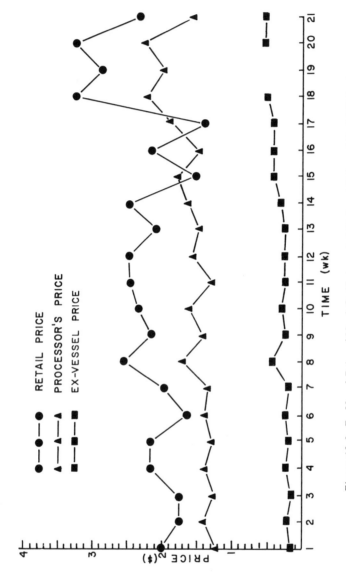

Figure 12.3. Ex-Vessel Price of Head-On/Eviscerated Cod and Prices of Cod Fillets at Processor and Retailer Levels.

quality was indeed high. More specifically, the most notable findings were as follows:

There was a complete absence of consumer complaints.
At the end of the experiment period, some consumers actively requested the graded and inspected product.
Consumers at times paid $1.00 more per pound for the product of assured quality than for products whose quality was not assured.
The quantity of fish returned unsold (returns) was greater in the beginning than at the end of the experiment.
Returns were generally lower when prices were low.
Returns were not consistently higher when prices were high.
Consumers tended to purchase familiar species over unfamiliar ones.
Retail managers were unanimous in their praise for the program.
The processor retained a keen interest in the program, even though at first sales amounted to only 200 pounds per week.

Fresh Fish: Experiment III

Experiment III dealt equally with the production, distribution and sale of fresh fish fillets under close control of the Gloucester Laboratory.

The systematic economic data collected and analyzed during this experiment helped to firmly establish that it was indeed economically feasible for the processor and the retailer to implement those quality assurance measures that have been discussed throughout this book. The gross profit margin was estimated to be $0.35 per pound for the processor and $0.51 per pound for the retailer. (As pointed out earlier, the gross profit margin should not be confused with the net profit margin, for which taxes and other overhead expenses must be calculated.)

Keep in mind that these estimates relate to an ideal—highly efficient—prototype producer and retailer, who might not be found in reality. Even so, the conclusion can still be reached that it is economically feasible to engage in a program of seafood quality assurance.

Let us now review some of the more detailed findings on which these overall judgments were based.

Summary of Added Costs To Assure Quality. As can be seen from table 12.1, the additional cost for the prototype processor to produce high-quality seafood was determined to be about $0.17 per pound ($1.16 − $0.99 = $0.17). To obtain the estimate of the cost differential for the program as a whole, this figure was reduced by about $0.07, an amount representing the cost reduction per pound experienced by the retailer, who no longer had to cut and package the fish in his store. Thus, for the

Table 12.1 Cost Estimates for Producing 1 Pound of Fresh Fillets of Assured and Not Assured Quality With Production Capacity of 10,000 Pounds Per Day

Assured Quality*		Quality Not Assured	
Cost Category	$ per lb.	Cost Category	$ per lb.
Ex-vessel price	$0.290	Ex-vessel price	$0.29
Gurry	0.630**	Gurry	0.55
Master case	0.080	Wooden boxes	0.03
Tray, pad, film	0.057	Tin cans, wax paper	0.04
Specialized filleting	0.040	Filleting	0.03
Specialized packaging	0.040	Random packing	0.03
Wrapping machine	0.001		—
Inspection	0.006		—
Transportation	0.020	Transportation	0.02
Overhead		Overhead	
Total	$1.16	Total	$0.99

*The added cost to produce fillets of assured quality was shown to be $1.16 − $0.99 = $0.17 per pound. This cost included prepackaging, which otherwise would have had to be done by the retailer. Thus the retailer saved this cost, which was estimated to be about $0.07 per pound. The added cost to assure quality for the program as a whole was therefore estimated to be about $0.10 per pound.
**The higher cost of the gurry for the assured quality fillet was due to the greater amount of trimming required.

program as a whole, the additional cost to produce and distribute fresh seafood of assured quality was estimated to be about $0.10 per pound. This estimate was confirmed numerous times by numerous sources. It must also be stressed that this cost might have been reduced even more if, as recommended, all fish that was not sold as fresh was taken out of the market and frozen while it was still of high quality.

Thus, even though the quality assurance program adds approximately $0.10 per pound to existing production and distribution costs, adoption of all recommendations can ideally lead to the production and distribution of seafood of assured quality at the same cost as seafood whose quality is not assured. This conclusion should not be surprising. The ultimate purpose of technological improvements is to lower costs— or at least to keep them stable—by increasing efficiency and reducing waste of resources.

Gross Profit Margins. The gross profit margin was estimated to be $0.35 per pound for the prototype processor and $0.51 per pound for the prototype retailer. These estimates provided only an indirect indication of the economic feasibility of the overall program. Since those profit margins ultimately had to be reduced by an undetermined amount spent for overhead costs, they gave no indisputable evidence that both economic operators—or, indeed, other economic operators—would be left with a specific profit margin. More direct evidence of the economic feasibility of the program was found elsewhere.

Proof of Economic Feasibility. While it did cost about $0.10 more per pound to produce and distribute high-quality fresh seafood, as can be seen from table 12.2, the consumer was ready to pay an average of $0.40 more per pound for fillets of assured quality. These two findings provided the needed evidence that the program offered a higher margin of profit to sellers of assured quality products than to sellers of products whose quality was not assured. Indeed, while the difference between the two sets of figures varied from experiment to experiment, for the high-quality products, it was always consistently well above the added costs of production and distribution.

In other words, it was found that, while it cost $0.10 less to produce fillets whose quality was not assured, their selling price to the consumer was also $0.40 less on the average. Hence, the differential on the sale of assured quality fillets was $0.30 per pound—the industry obtained $0.30 more for fillets whose quality was assured than for fillets whose quality was not assured. Since the bulk of production and distribution costs was the same for both types of products—there are costs in producing and selling products of any quality level—what is important here is the profit differential.

Many retailers and processors have confirmed these findings, not only informally to the experimenters, but more significantly by adopting the program. Sales of assured quality products increased considerably in a short time. For instance, estimates provided by the retailer indicated that sales for the chain as a whole—as distinguished from sales in the test stores—had increased by about 20 percent during this experiment (Machiaverna 1977); and another source reported that in a similar program of assured quality instituted by a twenty-six-unit chain

Table 12.2 Average Retail Prices for Fresh Fillets

Assured Quality		*Quality Not Assured*	
Species	$ per lb.	Species	$ per lb.
Cod	$2.25	Cod	$1.78
Flounder	2.85	Flounder	2.57
Haddock	2.73	Haddock	2.04
Ocean Perch	2.12	Ocean Perch	1.97
Pollock	1.65	Pollock	1.25
Overall average	2.32	Overall average	1.92

in Arizona, sales increased by 67 percent over a period of seven months (Zwiebach 1978). Indeed, we shall see that the program of assured quality—assisted by a large number of concurrent factors—gradually became part of a national trend of such importance that it induced many commentators to speak of a full-scale revolution in seafood marketing (Anonymous 1981). More pointedly still, it was said that "once a penance, fish is now posh pleasure" (Sheraton 1985, 92).

OTHER FINDINGS

Let us now briefly examine three other significant findings that were detected during this third experiment: a longer shelf life for fish of assured quality; a larger price fluctuation from week to week for fish than for meat products; and a wider spread between minimum and maximum prices for fish than for other meat products.

Fresh fish fillets delivered to the retailer in accordance with the recommendations of the quality assurance program lasted in a state of high quality for an average of five days, rather than the customary three days. In other words, by following the program's recommendations, the participants allowed the retailer—and ultimately the consumer—to handle a product that remained at a high-quality level for two extra days. Hence, the retailer had a greater assurance that he could sell his entire stock—and sell it while it was still of high quality.

With the exception of whiting, the retail—as well as ex-vessel and processor—price of seafood products under observation varied considerably from week to week. (The price stability for whiting was explained by its lack of popularity at the time and its availability at low and stable ex-vessel prices.) These fluctuations were not observed for other meat products.

The spread (difference) between the minimum and maximum retail price of seafood products was quite large—from $0.90 for pollock to $1.90 for cod—and it was larger than the spread for a selected group of meat and chicken products (see figure 12.4).

Of course, there are innumerable reasons—weather, species migration and restrictive regulations, to name a few—for the lack of price stability and large price spreads, two rather common factors in the seafood market. There are also innumerable reasons for the tendency of the price of meat and poultry to be generally much more stable than the price of seafood—most notably the fact that animals are domesticated allowing for nearly total control of operations from birth to slaughter. Hence prices for meat and poultry vary less from week to week, and when they do vary, they tend to swing less radically from a low to a high point. Erratic prices, coupled with erratic supplies, certainly

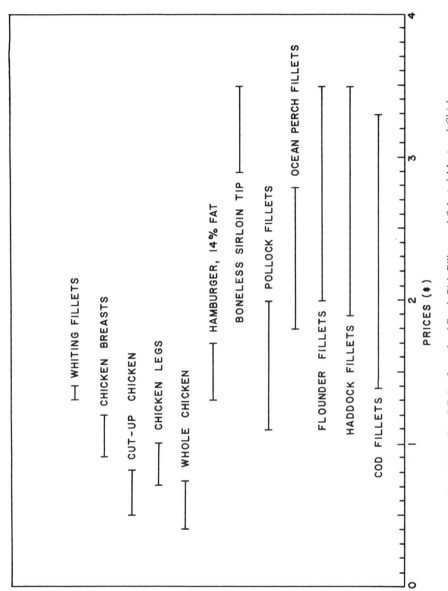

Figure 12.4. *Retail Price Spreads of Fresh Fish Fillets and Selected Meat and Chicken Products.*

do not bode well for the long-term health of the seafood industry, but these erratic trends are not unalterable. Indeed, smoothing erratic supply and price trends is—and should be—one of the major purposes of acquaculture and the frozen seafood market.

Some Effects of the Fresh Fish Experiment

The pilot project helped establish a set of trends of great significance. First of all, it proved that quality assurance is a technologically feasible program. It also helped to establish the recognition of the selling power of quality assurance to such an extent that, even without further assistance from the federal government, the study was often replicated and its recommendations became firmly rooted in the industry. Three early examples of duplication were the assured quality programs established by the state of Maine, the city of New Bedford, Massachusetts, and the government of Canada. Although, as with any endeavor, we are not convinced that the efficacy of these and subsequent programs could not be improved, there is no question that snags in the program will be eliminated only by trial and error.

Another trend originating from this experiment concerns the inspection and grading of fresh fish fillets. Prior to this pilot project, there was no official inspection and grading—indeed, no standard to permit the inspection and grading—of fresh (unfrozen) fish fillets. Such standards and practice had existed for years only for frozen fish fillets. During this experiment, the quantity of fresh fish fillets under government inspection grew from a few hundred pounds to several million pounds per year, and the program spread throughout the northeastern United States and beyond.

Other trends that originated during the course of this experiment include the prepackaging of fresh fish at the processor level (this practice, already common for frozen fish, eliminated a major source of disturbance at the retail level and favored the introduction of fresh fish into supermarkets); transportation of fresh fish in insulated containers; and the use of pull dates as an attempt to preclude the sale of fillets of less-than-high quality (also common throughout the food industry but not in the retailing of fresh fish).

Experiment IV: Frozen Seafood

Experiment IV observed the production, distribution and sale of frozen fish fillets focusing on the processor and the retailer. The Gloucester Laboratory had no control over these operations.

We will first analyze the findings regarding the operation of the processor and then those of the retailer.

Production Costs. Processor costs of raw material and sales prices are not given in detail here to avoid disclosure of proprietary information, but they were made available to the study. (Ex-vessel prices, a matter of public information, could not be used in this experiment because the processor bought fish that had been filleted by other processors.)

Overall unit cost for the production of frozen fillets whose quality was expected to be assured was found to be $1.74 per pound (no overhead costs were included in this estimate). Of special importance was the particularly high cost of raw material that had to be ascribed not only to the organizational structure of the business itself (the processor participating in the study was buying already filleted, rather than whole, fish), but also to the exceptional weather conditions prevailing when the experiment was initiated: In December 1980 and January 1981, New England harbors were ice-bound and almost no fish could be landed in those two months; in March and April, exceptionally high winds prevailed. Whatever fish was landed was sold at premium prices.

The cost differential of producing frozen fillets for the experiment was estimated by the processor—and not independently by the study—to be approximately $0.10 per pound. It must be stressed that this finding agrees with the cost differential of producing fresh fish fillets of assured quality as described earlier, even though the conditions of this study were different: The processor bought fillets already cut, and the retailer had no comparable cost savings.

Processor Markups and Gross Profit Margin. As can be seen in detail in table 12.3, processor markups varied from a low of $0.42 per pound to a high of $0.96 per pound. Taking into account quantities sold for each species, the overall weighted average markup (the difference between the buying cost and the selling price) was $0.78 per pound.

The gross profit margin for the prototype processor was estimated to be approximately $0.50 per pound.

Table 12.3   Processor Markups for Frozen Fillets (dollars per pound)

Species	January	February	March	April	May	June
Cod	0.732	0.880	—	0.762	0.60	0.70
Haddock	0.596	0.957	—	—	0.54	—
Pollock	0.550	0.419	0.459	0.551	0.55	—
Ocean Perch	0.870	0.762	0.579	0.943	—	—
Overall Weighted Average:	0.78					

Retail Sales. One of the most interesting phenomena was the break-down of sales by store, reported in table 12.4. Stores highlighted there are the five in which records were collected for the longest period of time and that were organized by geographic location, as shown in figure 12.2. It must be pointed out that sales for each store included fresh and frozen fillets of both assured and not assured quality. By correlating data in table 12.4 with those in figure 12.2, it was found that sales of frozen fillets of assured quality were the lowest—3 percent of total fish sales for the store—in the poorest area of downtown Albany (store #6); they were average—14 and 16 percent of total fish sales for the stores— in the high-income areas of uptown and suburban Schenectady (stores #2 and 4); and they were the highest—30 and 32 percent of total fish sales for the stores—in the middle-income areas of the western periph-ery of Schenectady and the northern periphery of Albany (stores #5 and 10).

The most intriguing total sales figures were those for the first nine weeks of the experiment. They showed that in this short time, retail sales for frozen fillets of assured quality grew consistently from 5 to 33 percent of total fish sales for the participating retail stores, and they grew to these levels at a very fast rate: In one store, sales for the fillets reached 16 percent of total fish sales during the first three weeks of the experiment.

After these encouraging signs, however, in subsequent weeks there was a consistent decline in sales, and it was evidently important to ana-lyze the reasons for this. We were compelled to exclude the potentially negative effect of high retail prices on sales, not only because prices varied too widely from species to species and week to week, but espe-

Table 12.4   Total Retail Sales of Frozen Fillets by Store (in pounds)

	*Store Number**				
Seafood Category	2	4	5	6	10
Frozen assured quality	663	478	1391	105	733
Frozen quality not assured	251	623	710	635	751
Fresh	3800	1814	2515	2321	834
Total	4714	2915	4616	3061	2318
Percentages					
Frozen assured quality	14%	16%	30%	3%	32%
Frozen quality not assured	5%	22%	15%	21%	32%
Fresh	81%	62%	55%	76%	36%

*Store number 10 remained in the study for 11 weeks, all other stores for 14 weeks.

cially because sales tended to be higher when prices were higher (notice in table 12.5 the higher prices for the first weeks). This evidence led us to undertake a different analysis. It was assumed that "quality" was the major explanatory variable in determining the decrease in the sales. Samples of the product were collected at various stages in their production and distribution chain and evaluated on a scale of 1 to 9, in accordance with appearance, odor, flavor, and texture (see chapters 4 and 5). As an overall result, a regression line of quality scores was obtained by aggregating the scores for all quality attributes and all species (Mendelsohn 1982). Upon review of the relationship between the sales trend and this regression line, a rather close correlation between the two became quite apparent. Even though the samples' scores never showed very high quality, it was still possible to observe that when scores were at the highest quality level found in this experiment—specifically, during the first weeks of observation—the sales trend was clearly upward. When the overall quality scores approached the lowest values, the sales trend was almost consistently downward.

This correlation seems to verify the validity of the hypothesis under which the entire study was conducted: High quality produces sales; low quality does not.

Some Causes of Quality Deterioration. Needless to say, in a strictly enforced quality assurance program no score should ever be below the 7 level (see chapter 7). In the present experiment, there were such scores because, for a variety of reasons, there was no control by the Gloucester Laboratory over various policies followed by the participants in the

Table 12.5. Retail Prices by Week: Frozen Fillets (dollars per pound)

Week	Cod	Haddock	Pollock	Flounder	Ocean Perch
1	3.49	3.49	2.09	3.89	3.69
2	3.49	3.49	2.09	3.89	3.69
3	—	—	—	—	—
4	—	2.99	—	2.99	—
5	2.99	2.99	2.09	3.89	3.69
6	3.19	3.29	1.99	3.89	2.97
7	3.19	2.98	1.99	3.89	2.98
8	—	—	—	—	—
9	3.19	3.29	2.09	3.89	3.69
10	3.19	3.29	2.09	3.89	3.69
11	2.69	3.29	2.09	3.89	3.69
12	2.69	2.89	2.09	3.89	3.69
13	—	—	—	—	—
14	—	—	2.09	—	—

experiment. In fact, without attempting to repeat here what has been stated throughout this book on the causes of deterioration of seafood quality, at least three related issues that belong to the business/economic field need to be mentioned. From the observation of production codes and discussions with the processor and the retailer participating in the experiment, it became apparent that the quality degradation that occurred in a number of supermarkets in a relatively short time was not only due to stocking of open-top display bins above the load line and to nonobservance of the first in, first out rule (hence, the importance of the experimental display case described in chapter 10 was fully confirmed), but also to initial overbuying (which led to overstocking of certain bins), to low turnover of the product, and to erratic prices.

The first two business issues are closely interrelated, in the sense that high initial volumes can clearly determine low turnover and low turnover breeds poor quality. (Needless to say, while quick turnover, assuming high initial quality, is an automatic quality controller, erring on the side of underbuying has the effect of slowing down the exploration of any latent demand in the market.) Those two causes—overbuying and low turnover—are also related to a third one, better discussed separately: pricing policy.

Pricing Policy. From the full range of data available to the study, only a small portion of which is reproduced here, it was clear that the supermarket chain participating in the study was in search of the best price level for each species of this essentially new product. As can be seen in table 12.5, it left only the price of fillets of flounder at the constant price of $3.89 per pound. It changed the price of pollock the least, but that of cod, haddock, and ocean perch considerably. It might be that this experimentation with prices had an undeterminable negative effect on the consumer. Was the product ever considered "overpriced" or "underpriced"? Was the linkage between high quality and high price ever broken in the mind of the consumer? The answers to these questions might provide explanations for the low turnover, the ultimate quality degradation of the product, and the declining sales in the last weeks of the experiment.

Some Relationships between Prices and Sales. While it is possible that the changing pricing policy had a negative short-term impact on sales, in the long run, it also contributed to the disclosure of three basic characteristics concerning the pricing of frozen fish fillets of assured quality. First, it appears that these fillets can be sold at considerably higher prices than fillets whose quality is not assured: With the exception of pollock, for which the difference was still about $0.70 to $0.80 per pound, for all other species, their price was consistently $1.00

to $2.00 higher than the price of frozen fillets whose quality was not assured. Second, the price of frozen fillets whose quality was supposed to be assured was generally as high as the price of fresh fillets. Thus, an assumed major deterrent to the production of high-quality frozen seafood, namely the expected higher production cost and lower sales price than fresh fish of comparable quality, appears to be without foundation: The equal or slightly lower price of frozen fillets is compensated for by lower spoilage levels. Third, high-quality frozen fillets seemed to sell more briskly at higher than at lower prices.

This last relationship is contrary to what generally happens with most products: Higher prices are supposed to dampen sales. It must be pointed out, however, that the above characteristics were all indications that, indeed, the product was at first perceived as being of "high quality," a fact confirmed by evidence gathered outside the confines of this study. For example, Pier 12, the brand name of the product under observation during this experiment, was reported to be "the fastest moving frozen brand" by Dave Conner, the seafood coordinator of Byerly's St. Louis Park supermarket in Minnesota. "'It outsells any of the frozen fish, I would say, 10 to 1,' Conner says. 'There is no frost buildup, no freezer burn and no shrinkage. It is a high-quality product'" (Cole 1981, 31–32). More important, perhaps, given this perception of high quality, is that the initial high price appears to have encouraged rather than deterred sales.

The willingness of the consumer to pay high prices can be explained not only by such sociological factors as "status symbol" or "conspicuous consumption," but also by the fact that it is wiser to spend more money for a usable product than less money for one that is worthless.

Retailer Markup and Gross Profit Margin. The retailer markup (the difference between the buying cost and the selling price) was found to be $0.88 per pound. Assuming that retail costs—including labor costs, costs of refrigeration, discards, etc.—were in the order of $0.20 per pound, it was possible to conclude that the retailer's gross profit margin was about $0.68 per pound. It should be stressed again that this is a broad estimate, and that it must be considered a gross profit margin because it does not even attempt to estimate overhead costs.

Economic Feasibility. As seen above, the processor and the retailer obtained a safe profit margin: $0.50 and $0.68 cents per pound, respectively. In addition, while production costs to assure quality were estimated to be about $0.10 per pound, the consumer was ready to pay $1.00 to $2.00 more for those fillets than for fillets whose quality was not assured. There seemed to be no question, then, that it was economically feasible to assure the consumer of the high quality of frozen fillets. And yet, from direct as well as indirect evidence, after its initial high promise, this experiment cannot be said to have been a success in either the short or the long run. The recommendations made have not taken root in the industry. There is no immediate evidence that the quality of frozen fillets in supermarkets has considerably improved. There is not even a faint indication of the long-range effects of high-quality domestic production over imports. And the reason seems to be evident: The results from Experiment IV failed to provide proof of the feasibility of consistent delivery of assured high-quality products to the consumer.

Technological Feasibility. Due to the lack of strict and consistent control over product quality throughout the chain of production and distribution, the results of Experiment IV demonstrated only sporadically and fitfully that it is indeed within the scope of current technology to assure the quality of frozen seafood to consumers. With more effective controls, it is expected that future experiments will provide proof of this feasibility on a continuing basis. In fact, the key deficiency in this experiment seemed to involve planning, coordination, and implementation of the required technological measures rather than the applicability of the measures themselves.

CONCLUSIONS

On the basis of the preceding findings, it is now possible to give some answers to the fundamental questions asked in this collaborative industry/government study in relation to seafood sold in supermarket stores:

1. It is technologically feasible to assure the consumer high-quality fresh seafood.
2. Experiment IV did not provide continuing proof that it is technologically feasible to assure the consumer high-quality frozen seafood, but did not disprove such a theoretical feasibility either. In a sporadic and fitful way, the product delivered to the consumer was indeed of a very high quality, but on the whole the quality was generally poor.
3. It did cost more to produce and distribute such fresh and frozen seafood, but only about $0.10 per pound more.
4. The consumer was ready to pay a higher price for such high-quality seafood—an average of $0.40 more per pound for fresh seafood and up to $1.00 and even $2.00 more per pound for frozen seafood of assured quality—thus giving indisputable proof of economic feasibility of the program.

5. If an organic program of the sale of fresh and frozen high-quality products were persistently implemented, not only might it have a reasonable probability of success in solving the very core of the problem associated with the sale of frozen fish fillets in the United States, namely the low price, quality, and reputation of frozen fish fillets; such a program might also begin to exploit the vast potential that exists in the frozen fish market in the United States and perhaps even start replacing some of the imports.

In brief, within the confines of this study, the question as to whether it is possible to sell high-quality seafood products can be given a qualified positive answer. And, given the existence of a wide socioeconomic spectrum of customers, it is possible to add that limited quantities of high-quality fish seem to be selling well at almost any price. Selling large amounts at almost any price is another matter, and this is discussed briefly in the next chapter.

Finally, it must be stressed that programs that keep the production and distribution of fresh fish separate from frozen fish will be less efficient than if the two programs were combined. The purpose of either program is to preserve as much fish as possible within high standards of quality. An ideal program of fish production and distribution would be one in which all fish that could be sold fresh should be, and the rest should be frozen while in a high-quality condition.

REFERENCES

Anon. 1961. Frozen fried fish sticks. *Consumer Reports* 26(2):80–83.

Anon. 1981. Seafoods: 'Seafood illiteracy,' trend to fresh at root of frozen fish sales lag: Safeway head. *Quick Frozen Foods* 43(12):51–54, 60.

Cole, J. B. 1981. Midwest retail: At Beyerly's, seafood pays. *Pacific Packers Report* 79(2):28–33.

Gorga, C., J. D. Kaylor, J. H. Carver, J. M. Mendelsohn, and L. J. Ronsivalli. 1979. The economic feasibility of assuring U.S. Grade A quality of fresh seafood to the consumer. *Mar. Fish. Rev.* 41(7):20–27.

Gorga, C., and L. J. Ronsivalli. 1982. International awareness for quality seafoods: A survey article. *Mar. Fish. Rev.* 44(2):11–16.

Gorga, C., and L. J. Ronsivalli. 1981. The importance of the U.S. seafood industry. *Seafood America* 1(7):26–27, 34.

Gorga, C., B. L. Tinker, D. Dyer, and J. M. Mendelsohn. 1982. Frozen Seafoods: The Economic Feasibility of Quality Assurance to the Consumer. *Mar. Fish. Rev.* 44(11):1–10.

Machiaverna, A. 1977. Grade A labels boost fish sales by 20%. *Supermarketing* 32(7):1.

Mendelsohn, J. M. 1982. U.S. Grade A Frozen Fish Program: Technological Report. In-House Report, National Marine Fisheries Service, Gloucester Laboratory, Gloucester, MA, 01930-2599.

Nickerson, J. T. R., and L. J. Ronsivalli. 1979. High quality frozen seafoods: The need and the potential in the United States. *Mar. Fish. Rev.* 41(4):1–7.

Ronsivalli, L. J. 1974. A study to determine the effect of assured quality of fish on its sales volume. An internal proposal. National Marine Fisheries Service, Gloucester Laboratory, Gloucester, MA 01930-2599.

Ronsivalli, L. J., C. Gorga, J. D. Kaylor, and J. H. Carver. 1978. A concept for assuring the quality of seafoods to the consumer. *Mar. Fish. Rev.* 40(1):1–4.

Ronsivalli, L. J., J. D. Kaylor, P. J. McKay, and C. Gorga. 1981. The impact of the assurance of high quality of seafoods at point of sale. *Mar. Fish. Rev.* 43(2):22–24.

Ronsivalli, L. J. 1981. U.S. seafood industry's big opportunity—quality assurance. *Seafood America* 1(9):24–28.

Sheraton, M. 1985. Just name your poisson: Once a penance, fish is now posh pleasure. *Time*, February 18, p. 92.

U.S. Department of Commerce. 1981. *Fisheries of the United States, 1980.*

Zwiebach, E. 1978. Basha's flying fish plan freshens section's sales. *Supermarket News* 27(18):54.

CHAPTER

13

The Potential Economic
Impact of Quality
Assurance

There are various methods through which one can attempt to measure
the impact of assuring the quality of seafood to the consumer. The simplest and most direct is to determine the growth rate and the value of
sales carried out under the banner of quality assurance. As we shall
see, this measurement takes into account all such sales—not just those
observed directly through the study undertaken by the Gloucester Laboratory (Gorga et al. 1978). This method can be defined as a straightforward study of sales.

A second and more thorough method consists of an attempt to determine the base—specifically, the current values concerning the seafood
industry as a whole—and then to assess the impact that quality assurance has had on this base. This method can be defined as a study of
supplies in the context of the industry in which they are generated.

A third, and the most comprehensive, method involves an estimate
of the impact of not only quality assurance per se, but quality assurance
in the context of an increase in the seafood supply, which, as predicted
in chapter 2, would most certainly be in evidence if the seafood was
indeed of high quality. This method can be defined as a study of the
pull of demand on supply.

FIRST METHOD OF ANALYSIS: STUDY OF SALES

In the previous chapter we saw that while production and distribution
costs increased by ten cents per pound, consumers were consistently

willing to pay substantially higher prices for fresh and frozen seafood whose quality was assured. The program was therefore financially feasible, but what was its scope? How large was this market? How large can it be expected to become? The questions are of general importance because, as pointed out in the previous chapter, it is one thing to sell small quantities of any product at a relatively high price, but the real issue is whether one can sell large quantities at those prices.

Three Measurements

Within the confines of the first method of analysis proposed above, there are three measurements of sales. The first concerns the measurements of sales effected during the life of the experiments run by the Gloucester Laboratory in collaboration with the industry. In each experiment, total sales were never more than a few thousand pounds so that, especially when broken down by species and by store, those measurements provided rather inconclusive results. No clear-cut trends could be discerned. As pointed out in the previous chapter, a more useful second measurement was that reported by other sources and informally offered by the participants in the study. They showed substantial growth rates: One source reported a 20 percent growth, another a 67 percent growth over short periods of time.

After nearly a decade of efforts in quality assurance, the Gloucester Laboratory made a more comprehensive analysis of the industry's progress in order to arrive at some measure of the validity of the original hypothesis that lack of high quality in seafood is the principal deterrent to consumption. Thus a third measurement was effected by measuring directly the sales of U.S. Grade A fresh products through figures provided by the USDC Inspection Service. The findings of this analysis exceeded expectations (Ronsivalli et al. 1981). As shown in figure 13.1, between 1976 and 1980 there was an exponential growth of additional sales (we shall return to this point) of fresh seafood brought about by the program of quality assurance. Within four years, the sales volume of U.S. Grade A fresh fillets had reached 11 million pounds per year with a value of about $30 million per year. The number of supermarket stores had increased to more than 1,100, the number of processors to more than 10, and the market area had expanded to include the fifteen northeastern states.

Impressive as these figures are, they do not portray the full picture. On the one hand, it must be reported that, presumably due mainly to the competition from imported high-quality seafood that does not necessarily bear the U.S. Grade A label, the sales volume of U.S. Grade A fresh fish fillets temporarily declined to 7.6 million pounds in 1984. One cannot give too much weight to this decline, however. First, by 1985 the

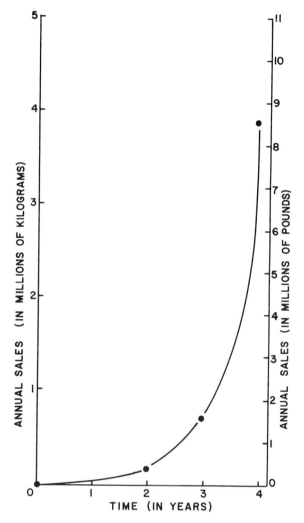

Figure 13.1. Effect of Quality Assurance on the Growth of Sales.

sales volume of this type of product had grown again to 12.2 million pounds. Second, as can be seen from table 13.1, the U.S. Grade A program of fresh and frozen seafood has a history of such considerable swings. Third, by 1981 the consistent technological assistance provided to the program by the Gloucester Laboratory had officially ended.

On the other hand, those figures do not report the full story for a whole set of other reasons. First, as pointed out earlier, the U.S. Grade A program is not identical to the assured quality program: Although not recommendable, there are sales of high-quality seafood not covered

Table 13.1 Edible Fishery Products Inspected (in thousand pounds)*

Year	U.S. Grade A**	Total†
1974	117,097	340,558
1975	138,230	623,403
1980	114,205	683,495
1981	109,612	624,760
1982	95,061	568,696
1983	91,934	566,720
1984	99,716	483,551
1985	117,720	442,772

*All figures refer to fresh and frozen products.
**Figures concerning the U.S. Grade A program were first reported in 1974.
†In addition to products carrying the U.S. Grade A label, the total is composed of products on which the USDC Inspection Service has affixed the "Packed Under Federal Inspection" (PUFI) mark, products that have been inspected but bear no mark as well as products that have been lot inspected.
Source: *Fisheries of the United States*, various years.

by the U.S. Grade A program. Second, while there is good justification that the figures for U.S. Grade A fresh fish were indicative of sales of assured quality, the same cannot be said for U.S. Grade A frozen fish fillets. Whatever sales of frozen fillets of assured quality have taken place, they cannot be separated from the official figures for fresh and frozen U.S. Grade A seafood as reported in table 13.1 and have not been recorded separately, either officially or unofficially. Thus whatever figures can be given here concerning sales of assured quality must be considered woefully underestimated. What is important is the overall trend. For about ten years seafood has been, and still is, one of the fastest-growing items sold in supermarkets. As Peter Redmayne put it, "red meat sales are on the run. Poultry has picked up some of the slack so far, but seafood is the protein with the most sales potential, food marketers agree" (Redmayne 1984, 40).

Are These Additional Sales?

It is admittedly difficult to determine whether the above figures represent additional sales, or substitute sales. If the latter, the net impact of the program would be near zero.

The direct measurement, namely the increase of the U.S. supply (domestic landings plus imports in equivalent round weight) from 7.4

billion pounds in 1976 to 9.2 billion pounds in 1985, is not the surest yardstick. Any number of causes—including lower prices, the growth of the population, or even an increase of ungraded imports—might have created that effect. (The possible effect of lower prices on sales can immediately be discarded. The index of retail prices shows that the price of fish had grown at a faster rate than the price of such substitute products as meat and poultry.) A different form of direct measurement would separate, and then compare, the yearly growth rates of graded and ungraded seafood, but this method is not entirely valid either at this stage, because the age and the base of the two types of sales are so completely different.

A better method—although indirect and certainly open to all sorts of questions, especially because not all seafood consumed is of the highest possible quality—consists of using per capita consumption figures. These at least offer the advantage that they eliminate the effect of population growth. Using this method one finds that whereas the *official* values of per capita consumption of seafood had oscillated between 10 and 12 pounds for the first seven decades of this century, it jumped from 12.9 pounds in 1976 to 13.6 pounds in 1984. In 1985 it had grown to 14.5 pounds. A complete record can be found in table 2.1 in chapter 2.

Any analysis of these figures reveals that it is obviously impossible to ascribe the growth in seafood sales to the quality assurance program alone. The increasing trend was already underway in the early 1970s: In 1973 per capita consumption had already reached 12.8 pounds, but fell back to 12.1 pounds in the following year. In addition to the discovery of the dietary significance of seafood, innumerable socioeconomic factors have undoubtedly intervened to foster that result. Some of them are easily identifiable: That we are going through a fad for fitness and even a minor revolution in the appreciation of the gustatory aspects of food in America is well known.

Whatever the specific effect of the quality assurance program on the growth of per capita consumption of seafood, this growth provides prima facie evidence that sales directly attributable to the program were additional and not substitute seafood sales. Total sales neither decreased nor remained constant; therefore, the quality assurance program did provide an undefinable incentive to the growth of seafood consumption in the United States.

A Cost/Benefit Analysis

Another indirect way to measure the effect of the quality assurance program run by the Gloucester Laboratory is to make a cost/benefit analysis. If one has reached a satisfying answer as to whether the sales ascribed to the program were indeed additional sales, the task can be immeasurably simplified by taking a straightforward measure of costs and benefits in relation to the nation as a whole. (The previous chapter can be read as a cost/benefit analysis of the industry as a whole.)

For instance, the cost to the federal government for the fresh fish segment of the study was $218,000. This expenditure covered the cost of returns, inspection, and transportation of the product under observation from the processor to the retailer. These expenditures were necessary not only to monitor the acceptance of the technology and the progress of the program as a whole, but especially to reduce to an acceptable level the financial risk to the participants (see chapter 12). These costs were a fixed expenditure. The benefits, however, still continue to increase. Simply taking into account the market value of additional sales, in 1980 the cost/benefit ratio was estimated to be 1:150 (Ronsivalli et al. 1981).

One can make these estimates as sophisticated and complex as one desires or needs simply by including in the analysis an increasingly more refined list of costs and benefits as well as by introducing an increasing amount of accuracy and precision in the measurement of each cost and each benefit. However, it is hard to escape the conclusion that a simple measure of the direct cost vs. value of sales of product at some reasonable point in time is indeed a fair indication of the return to the nation from expenditures of many of the federal programs discussed here.

A First Overall Estimate

However one studies the effects of the quality assurance program, the results do not fail to be impressive. The growth of sales for one supermarket chain, for example, were reported to be 20 percent and for another 67 percent. Sales for the program as a whole grew exponentially from a few hundred pounds to 11 million pounds in just five years. Measured indirectly, the sales due to quality assurance can be said to have contributed to the increase of 1.8 billion pounds in total sales and about 10 percent in per capita consumption over a period of about ten years.

Although the program undoubtedly contributed to the overall increase in seafood sales, however, the effect of a few other factors should at least be mentioned: the openness of American palates to new and exotic tastes; the passion for "fitness," as exemplified by the jogging fad; the discovery of healthful effects of seafood; the high quality of seafood sold in fast-food restaurant chains. All these and many others undoubtedly were the contributing factors for the increased consumption of seafood. But would this increase have been so dramatic if seafood had been of poor quality?

In any case, what stature does the program acquire when it is set in relation to the industry as a whole? To make this determination, we must become acquainted with the basic economics of the industry itself.

SECOND METHOD OF ANALYSIS: EFFECT ON THE ECONOMIC BASE OF THE INDUSTRY

No one disputes that seafood is important—if not vital—from a health and even dietary point of view. There are questions, however, as to the economic importance of the seafood industry in general and the seafood quality assurance program advocated here in particular. In fact, many people labor under the assumption that the U.S. fishery industry is insignificant. The reasons for this widespread belief are numerous, but perhaps the basic one is not difficult to detect. Since the seafood industry accounts for less than 1 percent of the Gross National Product (GNP) of the United States, it is assumed that it must be insignificant. A more complete and appropriate set of measurements and comparisons, however, offers a decidedly different view of the issue.

The Bare Facts

The facts concerning the U.S. seafood industry are summarized below, and, unless otherwise indicated, they refer to 1985, the last year for which official statistics are available at the time of this writing.

U.S. commercial landings of fishery products were 6.3 billion pounds, for a value of $2.3 billion. Of that total, 3.3 billion pounds were edible products valued at $2.2 billion and 3.0 billion pounds were commercial products valued at $128 million.

Total imports were valued at $6.7 billion: 2.8 billion pounds of edible products were valued at $4.1 billion and 530 million pounds of commercial products at $2.6 billion (corals and other valuable products are reported, not by weight, but by value only).

Since the total value of exports was $1.1 billion, the trade deficit was $5.6 billion.

(The relationship between domestic catch and imports merits a note of explanation. While domestic catch is reported in round weight, imports arrive in this country mostly in the form of fillets or wholly edible meats. To make the two sets of figures comparable, conversion factors are applied that transform figures for imports into round weight. This has been done in table 13.2, which shows the U.S. supply of human food from both sources.)

Table 13.2 U.S. Supply for Human Food (million pounds, round weight)*

	1970	1975	1980	1981	1982	1983	1984	1985
Domestic Catch	2.5	2.5	3.7	3.5	3.3	3.2	3.3	3.3
Imports	3.7	3.9	4.3	4.7	4.7	5.2	5.2	5.9
Total	6.2	6.4	8.0	8.2	8.0	8.4	8.5	9.2

*While domestic catch is in round weight, imports arrive in this country mostly in the form of fillets or wholly edible meats. To make the two sets of figures comparable, therefore, conversion factors are applied that transform figures for imports into round weight.

Source: *Fisheries of the United States*, various years.

In 1984, the U.S. seafood industry gave employment to 230,700 fishermen and to 109,623 people engaged in on-shore processing and wholesaling, for a total of 340,323 employees. The value added due to processing of both the domestic catch and imported products was a little over $7.5 billion in 1981, the last year for which this statistic has been published. This was the official estimate of the contribution by the U.S. fishery industry to the GNP, which then stood at about $837 billion: namely, nearly 0.9 percent. Another aggregate figure of interest is the total consumer expenditure for domestic and imported fishery products: $13.5 billion, again in 1981.

What do these figures mean? In order to obtain an answer to this question we need to put them in a larger context.

The Quantitative Importance

One measure of the importance of the U.S. fishing industry is obtained by setting it against the world fishing industry. In 1984, the United States landed 4.8 million metric tons of fishery products versus 82.8 million metric tons for the world as a whole. Although the U.S. catch represented only about 6 percent of the world catch, the U.S. fishing industry was still the fourth largest in the world. In fact, the first-ranking nation, Japan, landed only about 15 percent of the total world catch.

Another measure is obtained by setting the industry against the U.S. economy as a whole. True, the total value added by the fishery industry—or even its total consumer expenditures—represents less than 1 percent of the GNP, but this figure must be put in a more relevant context. It is misleading to use the GNP as a basis of the relative worth of any industry, because any single industry always represents only a small percentage of the GNP. The auto industry, for instance, usually considered extremely large, actually represented only 2.4 percent of the U.S. GNP in 1981.

Quite apart from overall national figures, it must also be recognized that the importance of the U.S. fishery industry varies from region to region. For instance, according to a May 23, 1979 report of the Department of Commerce Task Force on Fisheries Development, employment related directly to fisheries in Alaska in 1973 accounted for 19 percent of the total state employment and 7 percent of the gross state product. If related to individual cities and towns, the importance of the fishery industry might indeed be found to be so preponderant that it becomes vital even on economic terms alone.

A more significant gauge of the importance of the U.S. fishery industry can perhaps be found in the field of foreign trade. In 1984 the seafood trade deficit stood at $4.9 billion (it increased to $5.6 in 1985).

This figure becomes even more illuminating when you consider that in 1984 fishery products, when placed on the list of products accounting for the large current trade imbalance, were in a group of commodities that was overshadowed only by petroleum and petroleum products, various types of machinery (including automobiles), and clothing. In 1984 the value of imported seafood for food consumption was $3.7 billion, the highest figure in its commodity group; the next highest figures, $3.1 and $2.5 billion, were for coffee and vegetables, respectively.

Before entering any deeper into specific facts and figures, we must stress again that the fishery industry has unique and outstanding facets that must be borne in mind. Dollar values are an indiscriminate standard. They cannot distinguish financial from other values. The automobile industry, for instance, is undoubtedly important to society, but is it as important as food? In particular, is it as important as seafood, which, in addition to its dietary value, may have vital therapeutic properties? Air or water, for instance, do not have much economic value, but life is not possible without them. Seafood cannot be ranked that high on the list of man's basic needs, yet, as a source of special animal proteins, it is vital to survival.

Some of the most meaningful values are obtained when the U.S. fishery industry is compared with the meat and poultry industry, the other major sources of proteins. The fishery industry clearly occupies a second place in relation to these, but how far behind does it lie? In 1977, the figures for value added at the processing level alone were $7.4 billion for the meat industry versus $1.1 billion for fishery products, a ratio of 7:1.

This difference is substantial, but not as large as one would expect given the assumption that the fishery industry is insignificant. In fact, closer inspection reveals that the much-vaunted economic superiority of meat and poultry products compared to fishery products is still largely a myth, one sustained by an apparent huge discrepancy between per capita consumption of the relative products. The figures for 1983 were 176 pounds of meat products and 66 pounds of poultry products versus 13 pounds of fishery products. Much of the discrepancy, however, is simply due to the different methods of reporting the statistics. Meat and poultry products are given in carcass weight, but fishery products are reported in edible weight. Thus raw figures for fishery products have to be multiplied at least by three, or meat and poultry products divided at least by three, or perhaps by four, to make them comparable (fat is almost nonexistent in many of the popular species of fish, and bones are much heavier in meat products). It is evident that to avoid this common misunderstanding, there should be a standardization of the way food consumption rates are reported.

A note of great importance needs soon to be added: All the above figures relate exclusively to commercial fishing. If one were to add official estimates concerning recreational catches, many figures relating to the fishery industry would have to be increased by 30 percent. The entire exercise might not be valid, but one figure can certainly be increased by that amount: Per capita consumption of fishery products can safely be raised from thirteen pounds to about seventeen pounds. This figure still does not include the catches of freshwater and shellfish recreational fishermen, which would raise the per capita consumption figure for seafood to about twenty pounds.

Taking the above into consideration, comparable per capita consumption figures become from forty-five to sixty pounds for meat products, from fifteen to twenty pounds for poultry products, and from fifteen to twenty pounds for seafood products. To reduce the "softness" (lack of precision due to crude methods) of these estimates, one might also take into account the additional meat products derived from hunting, for which there seem to be no official estimates. What is known is only that in 1980 hunters were about half as numerous as sportfishermen, or seventeen million versus forty-two million people; in addition, hunters have much stricter restrictions (season, place, and even numerical quotas) than do fishermen. Assuming comparable productivity, one needs to raise the estimate for per capita consumption of meats, making it vary from sixty to eighty pounds.

The real gap between consumption of meat and fishery products, therefore, is considerably narrower than the one so frequently cited. The ratio expressing the gap is widely reported to be 18:1 for meats to fish and 6:1 for poultry to fish; using the more accurate analysis described above, the ratios expressing the real gap are more in the order of 4:1 for meats to fish and 1:1 for poultry to fish.

The discrepancy in per capita consumption between fishery and meat products is an indication of what can be achieved by the U.S. fishery industry. Stated very briefly, when the distribution of fishery products (which is still confined largely to coastal areas) becomes as widespread as it is for meat products—and when the quality of the two types of products becomes equally comparable—then, provided the supply can be increased to meet the demand, the per capita consumption of fishery products will increase and seafood imports might even decrease.

Finally, according to Martin, the editor of *Quick Frozen Foods*, a leading trade journal, in 1979 the value of frozen food sales climbed to almost $24 billion, with frozen fish sales for the first time surpassing $7 billion, thus representing 29 percent of this total. Although the increase in the value of frozen fish from the previous year was reported to be "*entirely* due to higher prices," the gain was nonetheless "so great that the category became the largest in dollar volume of them all, even exceeding prepared foods (Martin 1980, 14).

The Quantifiable Importance

The above figures relate to direct, already measured effects of the U.S. fishery industry. There are also innumerable indirect effects, however, that should be measured in order to have a more precise understanding of the true importance of the fishery industry. Although this measurement goes beyond the scope of this discussion, a mere—even though partial—list of those indirect effects should be mentioned. In addition to health effects, which have often been referred to in this book, three categories of such effects are of interest here: employment, income, and efficiency.

If precise calculations were made, how many jobs and how much income would be found to be generated—either in part or wholly— by the fishery industry for rope and twine producers, carpenters, and boat yard people, as well as real estate, insurance, or banking people, not to mention lawyers, public officials, and consultants? Such calculations would not be complete without adding figures related to the by-products of the fishing industry, from medicinals to cosmetics, from chicken feed to fish flour. Very positive, and quite surprising, finally are figures that can be included in a general category called efficiency. If the goal is to make proteins available to the consumer, fishery products are certainly the most efficient to produce.

The Qualitative Importance

Computers cannot, but the human mind can, integrate qualitative elements. No analysis would therefore be complete without mentioning some of those that are such an integral part of the fishery industry: tradition, aesthetics, and way of life.

Does not tradition contribute to the enrichment of every human being? American tradition, both literary and historical, is heavily tied to the fishing industry.

What is the value of looking at fishing vessels along the coastline of innumerable cities and towns? What is the value to the tourist industry of the view offered by such vessels on the horizon? Indeed, what would become of the appeal of such cities and towns if the fishing industry were to disappear?

"The life of a fisherman." To utter these words is to reduce such abstractions as ancient tradition and aesthetics to a living reality—a reality that includes hardship as an integral component.

A Second Overall Estimate

How important is the U.S. fishery industry, then? As the above discussion demonstrates, the fishery industry is a complex entity. As such, it is subject to differing interpretations. Ultimately judgments are always personal; they are generally formulated on the basis of one's own perspective and even one's own function in society.

Our position is clear. The myth of the insignificance of this industry must be dispelled once and for all. When the relevant facts are brought into their proper context, it seems impossible to reach any other conclusion but that the U.S. fishery industry is *quite* significant. In fact, the magnitude of its economic and social contribution to the nation and the potential for even greater contributions make the industry uniquely important.

Finally, measured against this base, how important is the quality assurance program? What is its economic impact? If one were to give a mechanical answer, one would simply calculate the net effect of the additional sales due to the program on the jobs, incomes, international trade, and all other aspects of the industry. Yet, even though the complexity of this study could be extended to any desired degree, it still might not help uncover the essential fact concerning the program. If faithfully carried out, the program improves the overall image of the entire industry. And, in an age in which aesthetic considerations are held quite high, an improved image is often translated into increased economic benefits.

THIRD METHOD OF ANALYSIS: PULL OF DEMAND ON SUPPLY

An increase in the level of quality of seafood available to consumers, especially in supermarkets, is likely to be translated into a call for an increased supply of seafood. This demand can possibly be satisfied by an increased level of imports but there are notable shortcomings to this policy. We therefore advocate that this demand be satisfied as much as possible by domestic fisheries. Indeed, we estimate that the local industry has much room for growth.

In particular, we predict that the five-point mission, discussed in chapter 2, consistently carried out with appropriate forms of federal leadership and assistance might in a relatively short span of time add about 4.0 billion pounds to the domestic supply of seafood. The detailed breakdown is as follows: aquaculture, 1.7 billion; fishery development, 750 million; preservation of temporary surpluses, 165 million; improving the efficiency of processing, 500 million; and savings due to quality assurance, 850 million.

In brief, current domestic production would be more than doubled. Any reservations concerning this estimate might be dispelled by the simple consideration that the future of aquaculture is nearly limitless. One simply needs to compare its potential with the current levels of poultry production to realize that the estimate is wholly within the realm of possibilities.

In order to analyze the effects of this third and final method, which gives an estimate of the full implementation of the quality assurance program, namely the potential delivery of the supply called for by an increase in the demand for quality seafood, one could rather simply double all the figures reported above, but certainly more sophisticated studies are warranted.

REFERENCES

Gorga, C., J. D. Kaylor, J. H. Carver, J. M. Mendelsohn, and L. J. Ronsivalli. May, 1978. The Technological and Economic Feasibility of Assuring Grade A Quality Of Seafoods. In-house report of the Gloucester Laboratory.

Gorga, C., and L. J. Ronsivalli. 1981. The importance of the U.S. fishing industry. *Seafood America* 1(7):26–27, 34.

Martin, S. 1980. Frozen food sales climb to $24 billion with fish & seafoods now top category. part I. *Quick Frozen Foods* 43(4):14–16.

Redmayne, P. 1984. Opportunity knocks. supermarkets lure shoppers with seafood. *Seafood Leader* 4(4):38–47, 64–67.

Ronsivalli, L. J., J. D. Kaylor, P. J. McKay, and C. Gorga. 1981. The impact of the assurance of high quality of seafoods at point of sale. *Mar. Fish. Rev.* 43(2):22–24.

Thompson, B. G. 1986. *Fisheries of the United States, 1985.* Washington, DC: U.S. Department of Commerce, National Oceanic and Atmospheric Administration, National Marine Fisheries Service.

U.S. Department of Commerce. 1979. *Toward a Partnership for the Development of the United States Commercial Fishing Industry.* Washington, DC: Fisheries Development Task Force, National Oceanic and Atmospheric Administration (May 23).

U.S. Department of Commerce. 1985. *Statistical Abstract of the United States: 1986.* Washington, DC: U.S. Bureau of the Census.

CHAPTER

14

Epilogue

In this book we have suggested that the United States—in a collaborative government/industry effort—embark upon a mission to assure the nation of an adequate supply of high-quality seafood. Specifically, we have maintained that it is not the fisherman, not the processor, not the retailer, but the consumer who is the ultimate judge of the level of quality of the product. The industry can give such assurance only if it keeps firmly in mind that, given the present state of technology, fresh seafood, if properly and constantly held at $0°$ C ($32°$ F), remains of high quality for only about nine days and frozen seafood, properly and constantly held at $-17.8°$ C ($0°$ F), remains in such a state for only about one year.

We have covered such topics as characteristics and methods of measurement of seafood quality, causes of its deterioration, and methods of prevention of such deterioration. We have particularly emphasized the differences between quality control and quality assurance, one basically being a set of scattered and uncoordinated measures particular to each firm, and the other a set of well-planned and well-executed, coordinated measures accepted across the industry. We have concluded that only the latter approach can deliver the guarantee that only products of high quality will reach the consumer—in particular when the channel of distribution is a supermarket.

In addition, we have concluded that it is quite economically and technologically feasible to produce and distribute high-quality fresh and frozen seafood (with some reservations about the need for added technology and/or education about the frozen seafood market), and that the assurance of high quality tends to increase the demand for seafood. Given the large number of favorable trends connected with seafood, which run the gamut from health to fashion, it is no wonder that there has been a veritable explosion in the rate of seafood consumption in this nation in the last few years. The question that remains to be asked is, Will it be domestic production or will it be imports to meet this demand?

We have suggested that the United States can and indeed should increase domestic production, through a variety of approaches, to try to satisfy that demand, and have proposed a twin set of goals: raise the quality of seafood, especially in supermarkets, and increase the domestic supply.

As seen in figure 14.1, a triple effect is then expected to ensue: Not only will the domestic productivity be raised and the current dependency on foreign imports be lowered, but also the health of the nation's population will improve.

Is there any better use for fish—this natural, renewable resource with which the United States has been so richly endowed?

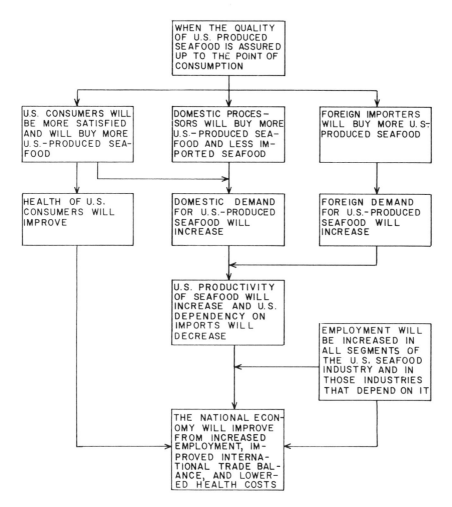

WHEN THE QUALITY OF U.S. PRODUCED SEAFOOD IS ASSURED UP TO THE POINT OF CONSUMPTION

U.S. CONSUMERS WILL BE MORE SATISFIED AND WILL BUY MORE U.S.-PRODUCED SEAFOOD

DOMESTIC PROCESSORS WILL BUY MORE U.S.-PRODUCED SEAFOOD AND LESS IMPORTED SEAFOOD

FOREIGN IMPORTERS WILL BUY MORE U.S. PRODUCED SEAFOOD

HEALTH OF U.S. CONSUMERS WILL IMPROVE

DOMESTIC DEMAND FOR U.S.-PRODUCED SEAFOOD WILL INCREASE

FOREIGN DEMAND FOR U.S.-PRODUCED SEAFOOD WILL INCREASE

U.S. PRODUCTIVITY OF SEAFOOD WILL INCREASE AND U.S. DEPENDENCY ON IMPORTS WILL DECREASE

EMPLOYMENT WILL BE INCREASED IN ALL SEGMENTS OF THE U.S. SEAFOOD INDUSTRY AND IN THOSE INDUSTRIES THAT DEPEND ON IT

THE NATIONAL ECONOMY WILL IMPROVE FROM INCREASED EMPLOYMENT, IMPROVED INTERNATIONAL TRADE BALANCE, AND LOWERED HEALTH COSTS

Figure 14.1. A Summary of Expected Effects of Quality Assurance of Seafood.

APPENDICES

1

A Few Selected
Comments on Lipids

The fats and oils discussed here are both formed by a reaction that takes place within biological systems. In the reaction, glycerol combines with one, two, or three molecules of compounds known as fatty acids, thus the terms monoglycerides, diglycerides, and triglycerides (see figure A1.1).

Monoglycerides and diglycerides are not common in nature, so we shall restrict our discussion to the triglycerides, of which most fats and oils are composed. In figure A1.1, the formula for fatty acid(s) is represented by the symbol R, which stands for the hydrocarbon chain portion of the fatty acid molecule. The second portion of this molecule is the carboxyl group (see figure A1.2). In order to write shorthand formulas for dissimilar fatty acids we add a subscript to the R in each formula. Accordingly, R_1COOH differs from R_2COOH, from R_3COOH, and from any fatty acids whose R has a subscript different from 1. Fatty acids with the same subscripts are the same (e.g., see the fatty acids in the simple triglyceride in figure A1.3).

The triglycerides vary in their properties, depending on the fatty acids in their composition. In some cases the glycerol is linked to three molecules of a particular fatty acid (called a simple triglyceride); in mixed triglycerides, which are the most common, the glycerol is linked to three molecules from three different fatty acids or from two different fatty acids (see figure A1.3). For the purposes of this discussion we

GLYCEROL + FATTY ACID → MONOGLYCERIDE + WATER

$$
\begin{array}{c}
H \\
| \\
H-C-OH \\
| \\
H-C-OH \\
| \\
H-C-OH \\
| \\
H
\end{array}
\;+\; RCOOH \longrightarrow
\begin{array}{c}
H \\
| \\
H-C-OH \\
| \quad\quad O \\
| \quad\quad \| \\
H-C-O-C-R \\
| \\
H-C-OH \\
| \\
H
\end{array}
\;+\; H_2O
$$

GLYCEROL · FATTY ACID · MONOGLYCERIDE · WATER

GLYCEROL + 2(FATTY ACIDS) → DIGLYCERIDE + 2 WATER

$$
\begin{array}{c}
H \\
| \\
H-C-OH \\
| \\
H-C-OH \\
| \\
H-C-OH \\
| \\
H
\end{array}
\;+\; 2(RCOOH) \longrightarrow
\begin{array}{c}
H \quad\quad O \\
| \quad\quad \| \\
H-C-O-C-R \\
| \quad\quad O \\
| \quad\quad \| \\
H-C-O-C-R \\
| \\
H-C-OH \\
| \\
H
\end{array}
\;+\; 2(H_2O)
$$

GLYCEROL · FATTY ACIDS · DIGLYCERIDE · WATER

GLYCEROL + 3(FATTY ACIDS) → TRIGLYCERIDE + 3 WATER

$$
\begin{array}{c}
H \\
| \\
H-C-OH \\
| \\
H-C-OH \\
| \\
H-C-OH \\
| \\
H
\end{array}
\;+\; 3(RCOOH) \longrightarrow
\begin{array}{c}
H \quad\quad O \\
| \quad\quad \| \\
H-C-O-C-R \\
| \quad\quad O \\
| \quad\quad \| \\
H-C-O-C-R \\
| \quad\quad O \\
| \quad\quad \| \\
H-C-O-C-R \\
| \\
H
\end{array}
\;+\; 3(H_2O)
$$

GLYCEROL · FATTY ACIDS · TRIGLYCERIDE · WATER

Figure A1.1. The Formation of Glycerides (Fats and Oils).

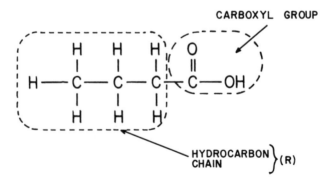

CARBOXYL GROUP

HYDROCARBON CHAIN ⎫(R)

Figure A1.2. Components of Butyric Acid, a Fatty Acid.

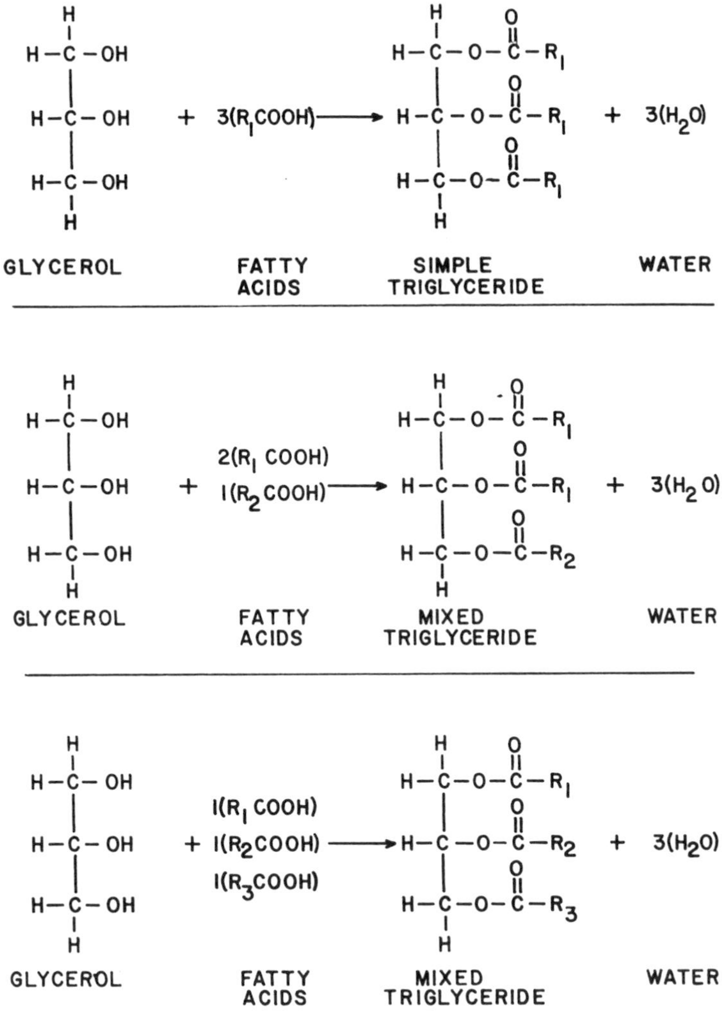

Figure A1.3. The Formation of Simple and Mixed Triglycerides.

shall not pursue further the nature of glycerol or the chemistry that combines it with fatty acids or of the proportion of fatty acids in each lipid molecule. Instead, we shall focus our attention on the fatty acids, for it is around these that we expect to show the outstanding contribution of seafood to the diet.

Fatty acids are made up of two parts: a carboxyl group (the name given to the [COOH] radical) and a hydrocarbon chain (a string of carbon atoms with hydrogen atoms attached to each). For example, look at the chemical structure for butyric acid in figure A1.2. In the formula, each C stands for one atom of the element carbon, each H for one atom of the element hydrogen, and each O for one atom of the element oxygen, and each line that connects the different atoms stands for a chemical bond. Note that each C atom is linked by four bonds to other atoms. Since each of the fatty acids contains a carboxyl group, they do not differ from each other in that respect. Where they do differ is in the length (sometimes the shape) of the hydrocarbon chain and in the degree of unsaturation within the chain. It is exactly because of these two differences that lipids are labeled either fats or oils, and it is the hydrocarbon chain that accounts for the reactivity of the lipid molecule and influences its overall character.

SATURATION, UNSATURATION, POLYUNSATURATION

The lengths of fatty acids in the lipids found in nature vary in an even-numbered sequence from four to twenty-six C atoms. Two smaller acids, formic acid (two C atoms) and propionic acid (three C atoms) are not normal components of natural lipids. Hence, butyric acid, shown in figure A1.2, is the smallest of the fatty acids encountered in lipid molecules. When each C atom in the hydrocarbon chain of a fatty acid (note that, here, we are excluding the C atom in the carboxyl group where a double bond is shown linking the C atom and the O atom, as in figure A1.2) is linked to at least two H atoms, there will be no double bonds between the C atoms. Such fatty acids are said to be saturated. Thus, butyric acid is a saturated fatty acid.

When there are two C atoms connected by a double bond in the hydrocarbon chain of a fatty acid (again excluding the carboxyl group), then the fatty acid is said to be unsaturated at that site, and the molecule will contain two H atoms less at the site of double bond. If there is more than one site of unsaturation in the hydrocarbon chain of a fatty acid, the fatty acid is said to be polyunsaturated. Figure A1.4 shows structural formulas of portions of the hydrocarbon chains of fatty acids to illustrate this point. It should be noted that there may be more than two sites of double bonding in polyunsaturated chains.

The melting points of the saturated fatty acids, as typically found in meats, usually are at temperatures higher than room temperature; therefore lipids that are made up of saturated fatty acids are generally in the solid state at room temperature. Unsaturated and polyunsaturated fatty acids in plants and seafood usually have melting points below room temperature and are therefore in the liquid state at room temperature.

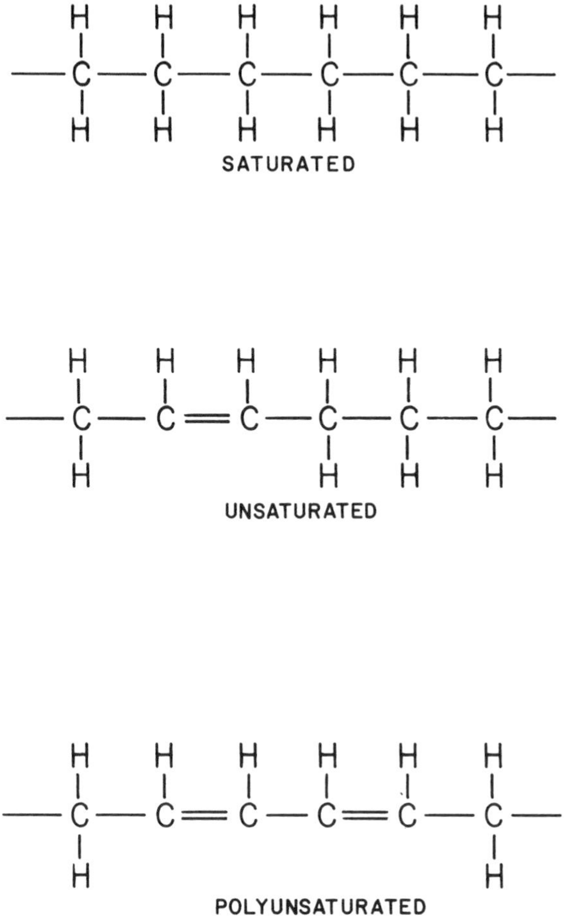

Figure A1.4. Structural Formulas of Saturated, Unsaturated, and Polyunsaturated Hydrocarbon Chains in Fatty Acids.

APPENDIX

2

Effects of Decreasing Temperature on the Physical State of Water in Fish Tissue

This appendix provides some insights into the effects of decreasing temperatures on the physical states of the water contained in fish and fish fillets. Pure water freezes at 0° C (32° F), but this freezing point is depressed when water contains dissolved substances such as salt. The water in biological (living) systems contains varying amounts of dissolved substances (even after death of the organism), therefore the water in these systems will not freeze at 0° C (32° F), but will begin to freeze at a lower temperature (about −0.9° C or 30.4° F). Moreover, whereas pure water freezes completely at 0° C (32° F), the water in biological systems freezes gradually as the temperature is lowered—but not completely—even after the temperature is reduced to levels that are much lower than those available in commercial or domestic freezers (see figure A2.1).

Thus, when the temperature of a seafood is progressively lowered, the physical transformation of the product to a rigidly frozen mass occurs gradually. The rate of ice formation within the fish tissue is fastest in the beginning and becomes progressively slower as the temperature is decreased at a linear rate. By the time the product temperature is lowered to about −3° C (26.6° F), so much ice has formed that the product becomes rigid and unyielding except to heavy pressure. At about −6° C (21.2° F), approximately 85 percent of the water in the fish has been frozen, and at this stage even heavy pressure will not deform

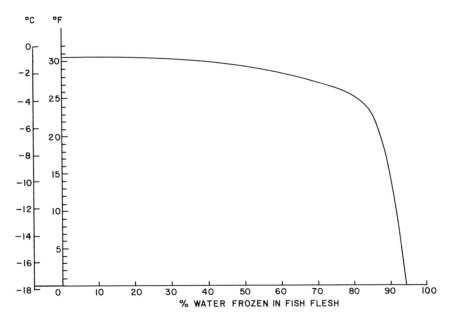

Figure A2.1. The Effect of Lowering the Temperature of Fish Flesh on the Physical State of Its Water Content.

the frozen tissue. Also, at this stage, the rate of increase in the amount of ice slows down considerably, and even at the coldest temperatures found in commercial freezers (which are lower than in domestic freezers and in the freezer display cases in retail markets), some of the water in the fish tissue remains unfrozen.

Bound water, the water that is an integral part of the chemical structure of tissue components (e.g., proteins), is reported to resist freezing at extremely low temperatures. In one series of experiments to measure the amount of water remaining unfrozen in seafoods, it was found that, using an indirect method to measure water activity, it was not until the temperature was lowered to about $-55°$ C ($-67°$ F) that all of the water had been frozen.

It can therefore be seen readily that while about 85 percent of the water is frozen as a result of a relatively small decrease in temperature (about 5° C or 9° F), it requires a dramatic ten-fold temperature decrease to freeze the remaining 15 percent of the water. Clearly, the water contained in "frozen fish" is never completely frozen.

3

Importance of the Precision and Accuracy of Measurement

The main requisite of a test for the quality of seafood is that it be reliable in its findings. Two vital elements of reliability are the precision and the accuracy of the testing method.

Precision. Precision can be defined as the fineness with which a measurement can be made. For example, weigh scales showing kilograms and grams (or pounds and ounces) are more precise measuring tools than those showing only kilogram (or pound) values: A 2.4 kg (5 lb, 4.6 oz) item can be weighed with greater reliability with the first set of scales than with the second. The latter scales would simply show that the weight of the item is between 2 and 3 kg (5 and 6 lb) but would not yield a value expressing the exact weight.

Using this definition of precision as a base, let us now go on to other technical terms used in the measurement of any quantity.

Greatest Possible Error. The greatest possible error is defined as the error that equals half the smallest increment of measurement. Also, the smallest increment of measurement determines the precision of measurement. Thus, the greatest possible error made using the kilogram (or pound) scales cited above is 0.5 kg (or 0.5 lb). On the other hand, the greatest possible error made using the kilogram/gram (or pound/ounce) scales cited above is 0.5 g (or 0.5 oz). The finer the precision of a measuring instrument, the smaller the error of measurement — and the greater the accuracy.

Relative Error. The relative error, also called the percentage error, is the ratio of the greatest possible error to the recorded measurement.

$$\text{Relative Error} = \frac{\text{Greatest Possible Error}}{\text{Recorded Measurement}}$$

Using the scales above to weigh an item having an actual weight of 2.4 kg (5 lb, 4.6 oz), we obtain the following values of the relative error.

1. Using the kilogram/gram scale, the relative error is

$$\frac{0.5 \text{ g}}{2.4 \text{ kg}} = \frac{0.5 \text{ g}}{2400 \text{ g}} = 0.0002 = 0.02\%.$$

2. Using the pound/ounce scale, the relative error is

$$\frac{0.5 \text{ oz}}{5 \text{ lb, 4.6 oz}} = \frac{0.5 \text{ oz}}{84.6 \text{ oz}} = 0.006 = 0.6\%.$$

3. Using the kilogram scale, the relative error is

$$\frac{0.5 \text{ kg}}{2.4 \text{ kg}} = 0.21 = 21\%.$$

4. Using the pound scale, the relative error is

$$\frac{0.5 \text{ lb}}{5 \text{ lb, 4.6 oz}} = \frac{0.5 \text{ lb}}{5.29 \text{ lb}} = 0.09 = 9\%.$$

Accuracy. Accuracy in measurement is attained only when the true value of what we are trying to measure is found exactly. In most cases, and particularly in cases involving natural phenomena, standard measuring instruments are just not fine enough (do not have the adequate precision) to make a strictly accurate measurement. For example, a flower stem 25.4001 cm (10.000039 in) long cannot be measured exactly with an ordinary centimeter scale, which has subdivisions of 1 mm (0.03937 in). To get an exact measurement, we could use sophisticated instrumentation such as the electron microscope, which allows us to obtain satisfactory measurements in microns (10^{-4} cm or 3.937×10^{-5} in). But even the electron microscope may not allow us to determine the exact size of a bacterium, which cannot be measured exactly except by using even greater precision. In practical terms, many of the measurements we make are considered to be accurate. In a strictly technical sense, few if any are accurate—and those that are, most likely are accurate by chance.

For any given measurement, the finer the precision of the instrument, the greater the accuracy. But accuracy is not synonymous with precision, nor is it dependent directly on precision. Accuracy is dependent on, and is expressed in terms of, the relative error. The lower the relative error, the greater the accuracy. This means that accuracy is proportional to the inverse of the relative error: accuracy α 1/relative error.

4

Spoilage Rates
of Seafood

Although it is known that the spoilage of fish is much more complicated than a simple chemical reaction, it is nevertheless possible to obtain a relatively reliable measure of the rate of spoilage in seafood. This is possible because regardless of the complexities of the number and types of reactions occurring simultaneously, one reaction limits the rate at which the entire process of spoilage or deterioration of quality occurs. The hypothesis of the existence of a single limiting reaction can be tested empirically. When a sufficient number of organoleptic tests of the spoilage rates of fish at different temperatures is made, a plot of the rates versus the storage temperatures yields a relatively linear relationship in the range 0° C to 8° C (32° F to 46° F). The linearity of the relationship does not appear to extend beyond this temperature range, which can be interpreted as evidence that other reactions are responsible for limiting the rate of spoilage outside this range.

Empirical studies have been conducted by research teams in Great Britain (Spencer and Baines 1964), Australia (James and Olley 1971), and the United States (Charm et al. 1972) working independently but aiming toward similar goals. The published results of the findings of these three efforts revealed striking similarities. Applying the Arrhenius equation, the English and Australian teams found the activation energy of the spoilage reaction to be in the range of 15,000 to 19,000 calories per mole, and the United States team found it within a much narrower range at slightly over 18,000 calories per mole. It is believed that the tighter range of activation energy values found by the U.S. team was due to a difference in the analytical procedure employed in the organoleptic tests, as described in chapter 5. Here it is relevant to discuss only one term of the Arrhenius equation. This equation, which is named after the famous Swedish chemist who is credited with having developed it, relates temperature and the rates of chemical reactions. It reads:

$$K = Ae^{-H/RT}$$

where, K equals the rate of reaction; A, the orientation factor; e, the mathematical constant; H, the energy of activation; R, the gas constant; and T, temperature (absolute scale).

The orientation factor represents the frequency of incidence when the reactive sites (moieties) of the molecules involved in the reaction are so aligned as to be able to react. This is an important point to consider, because it might explain why the spoilage rate of fish is not affected by the number or types of bacteria in the middle to later stages of spoilage (increased numbers of bacteria means an increase in the amount of enzymes as well). Thus, we theorize that once enzymes attain such a level that the reaction rate is constant, the introduction of additional enzymes will not increase the rate of spoilage because the enzyme level already present is enough to react with all the available reaction sites in the flesh substrate.

Another influencing factor in the spoilage rate is that spoilage is largely a surface phenomenon. This condition implies that both the bacteria that produce the spoilage enzymes and the reactions involving the enzymes require oxygen, and it accounts for the fact that large and/or round fish that have a relatively small surface-to-volume ratio appear to spoil at a slower rate than small and/or flat fish, which have a relatively high surface-to-volume ratio. It also accounts for the apparently faster rates of spoilage in chopped or minced fish, where the surface-to-volume ratio is made extremely large. This result should not be surprising, since the dependency upon oxygen by the bacteria responsible for the spoilage of seafood can readily be shown. For example, of four groups of pseudomonads found on seafood, the two highly active in the spoilage of seafood are aerobes.

REFERENCES

Charm, S.E., R.J. Learson, L.J. Ronsivalli, and M.S. Schwartz. 1972. Organoleptic technique predicts refrigeration shelf life of fish. *Food Technol.* 26 (7):65–68.

James, D.G., and J. Olley. 1971. Spoilage of shark. *Aust. Fish.* 30 (4):11–14.

Spencer, R., and C.R. Baines. 1964. The effect of temperature on the spoilage of wet white fish. *Food Technol.* 18: 769–773.

5

Formulas for the Determination of Gross Profit Margins and a Few Specific Costs

The formulas used for the determination of gross profit margins and a few specific costs in this study are standard. However, since they include factors that are particular to the seafood industry or to the study at hand, it might be useful to spell them out.

Gross Profit Margin for the Processor

$$PGPM = PP - (VP + G) - PC$$

where *PGPM* stands for processor gross profit margin, *PP* for processor selling price, *VP* for ex-vessel price, *G* for net price of gurry (see below for a fuller explanation of *VP* and *G*), and *PC* for processor operating costs. (Overhead costs were not available to the study.)

Gross Profit Margin for the Retailer

$$RGPM = RP - PP - RC$$

where *RGPM* stands for retailer gross profit margin, *RP* for retailer selling price, *PP* for processor selling price, and *RC* for retailer operating costs. (Overhead costs were not available to the study.)

Whenever necessary, average processor and retail selling prices were weighted with respect to quantities for each species.

Unit Cost of Raw Material

$$RM = (VP)(Lbs) \pm G$$

where RM stands for unit cost of raw material, VP for ex-vessel price, Lbs for number of pounds of whole fish bought, and G for net cost of gurry or waste products (which can either be sold at a small profit or disposed of at an additional loss, depending essentially on the availability of waste reduction plants to transform the gurry into fish meal or other products). G is calculated on the basis of the yields, namely the edible portion or fillet. The yield varies from species to species—as well as according to the desired level of quality. The U.S. Grade A level, for instance, requires additional trimming to improve the appearance of the fillet and/or eliminate the presence of bones and other defects. The fulfillment of these requirements is identified as specialized filleting. For the cost estimates, reported in the text (186, 189–90, table 12.1) the yields for assured and those for not assured quality fish were calculated as follows (the yields for fish whose quality was not assured are in parentheses): cod, 29 (37) percent; flounder, 28 (30) percent; haddock, 38 (41) percent; ocean perch, 27 (27) percent; pollock 36 (44) percent; whiting, 30 (34) percent.

An example might help clarify the major elements of this formula. A processor who produces fish fillets buys whole fish as his raw material, but sells only the edible portion or the fish fillet. If he buys from a fisherman, he buys at what is known as the ex-vessel price. This is the price set for each catch (or portion of the catch) on a per pound basis for whole fish. The ex-vessel price varies from species to species and usually day to day. For example, on a given day the ex-vessel price for cod might be \$.75 per pound. This means that 1000 pounds of whole, eviscerated cod will cost the processor \$750.00. But in the production of fillets, the 1000 pounds of whole eviscerated fish will yield, say for the sake of this discussion, about 333⅓ pounds of fillets. He now has 333⅓ pounds of fillets and 666⅔ pounds of gurry (bones, skins, heads, etc.). If he disposes of the gurry without any added cost to him, his adjusted cost for the 333⅓ pounds of usable raw material (the fillets) will remain \$750.00 but the per pound price will go up to \$2.25. If it costs him some money to dispose of the gurry, say \$50.00, then the cost for his raw materials will be \$750.000 + \$50.00 and the per pound cost, \$2.40. If he can sell the gurry to a manufacturer of fish meal, fertilizer, or other commodity for, say, \$.02 per pound, then his adjusted cost will be \$750.000 − (\$0.02 × 666⅔) or \$2.21 per pound of fillets. Thus, the unit cost of the raw material is basically equal to the ex-vessel price of the fish, and it may be more or less depending on how a processor disposes of the gurry.

Unit Cost of the Wrapping Machine

$$WM = \frac{(UP)(ALb)(T)}{C}$$

where *WM* stands for unit cost of wrapping machine, *UP* for units packaged per day, *ALb* for average weight per unit (independently determined through specific surveys), *T* for production time (specifically assumed to be five days per week, forty-eight weeks per year, for five years, the estimated life of the machine), and *C* for the cost of the machine.

Unit Cost of Labor Attending Wrapping Machine

$$LWM = \frac{(UPP)(N)(ALb)(T)}{W}$$

where *LWM* stands for the unit cost of labor attending wrapping machine, *UPP* for human capacity or units packaged per person per hour, *N* for number of attendants per machine, *ALb* for average weight per unit (independently determined through specific surveys), *T* for production time (specifically assumed to be seven-hour work day), and *W* for total labor cost per unit of time.

Inspection Cost per Unit Produced

$$I = \frac{(Lbs)(T)}{C(t)}$$

where *I* stands for inspection cost per unit produced, *Lbs* for total number of pounds that, in an efficient operation, could be (rather than were) inspected per unit of time, *T* for unit of production time, *C* for cost per unit of inspection time, and *t* for total inspection time.

A similar formula was developed to calculate transportation costs and the cost of returns, because the conditions were the same as in the case of inspection costs.

Index